D1441380

AIR CONDITIONING
AND
MECHANICAL TRADES
Preparing for the Contractor's License Examination

AIR CONDITIONING
AND
MECHANICAL TRADES
Preparing for the Contractor's License Examination

John Gladstone

VNR VAN NOSTRAND REINHOLD COMPANY
NEW YORK   CINCINNATI   TORONTO   LONDON   MELBOURNE

Van Nostrand Reinhold Company Regional Offices:
New York Cincinnati Chicago Millbrae Dallas
Van Nostrand Reinhold Company International Offices:
London Toronto Melbourne

Copyright © 1975 by John Gladstone
Library of Congress Catalog Card Number: 74-18258
ISBN: 0-442-22701-9

Manufactured in the United States of America

Published by Van Nostrand Reinhold Company
450 West 33rd Street, New York, N.Y. 10001

Published simultaneously in Canada by
Van Nostrand Reinhold Ltd.

15 14 13 12 11 10 9 8 7 6 5 4 3 2 1

Library of Congress Cataloging In Publication Data

Gladstone, John
    Air conditioning and mechanical trades.

    Includes bibliographies.
    1. Air conditioning —Examinations, questions, etc.
2. Mechanical engineering — Examinations, questions, etc.
I. Title
TH7687.5.G55   697.9'3   74-18258
ISBN 0-442-22701-9

PREFACE AND ACKNOWLEDGMENTS

Although candidates for professional engineers' licenses, civil service examinations, radio license, and other professions and trades, have for a long time had access to various study guides and review books of actual previous exams, a mantle of secrecy hangs over the examination material for master mechanical contractor and subcontractor trades. Whether this is owing to the newness of such examinations or to the professional jealousy and business secrecy that seems to plague the traditional contractor boards, it remains that no book on the subject has heretofore appeared.

Preparing this book I was continuously confronted with the problem of what material to include and what to exclude. Too much would make the book ponderous; too little might not fulfill the need. Being limited to a fixed number of pages, after much deliberation, I decided to omit questions dealing with controls, fire sprinklers, elevators and incinerators in favor of a little refresher on basic math, heating, electric and insulation. In addition, I included some important reference data in the Appendix not generally found in a book of this kind. A section of formulas and equations was dropped in the interest of space conservation. The important equations appear in their respective sections, e.g., psychrometric equations appear in the section 'Psychrometrics',etc.

The work is organized into nine sections. The Introduction is a discussion about examinations in general to familiarize the reader with the language, construction and psychology of tests. The second section gives actual questions and answers from seven different categories of mechanical subcontractor crafts. Usually, questions from all of these categories will be asked on the Master Mechanical Contractor Examination, but a candidate for say, Master of Sheetmetal and Ventilation need be responsible only for questions dealing in his subcategory. The third section covers national and local codes for heating, piping, ventilating, and air conditioning. The fourth section conducts a short treatise on some sticky problem-solving questions that are likely to appear on an examination. The fifth section discusses general safety and rigging. The sixth section prepares the candidate for questions dealing with business management, and accounting procedures. The seventh section offers an overview of contractor law including a thorough review of OSHA. The eighth is a brief review of basic math. The ninth gathers together some useful data on electrical, heating, and insulation as a source of reference for odd examination questions. Finally, the Appendix offers tables of conversion data and other important, hard-to-find reference material.

This book is largely a reorganization and presentation of existing information; only the smallest passages offer original ideas. For this reason I wish to acknowledge my indebtedness to the work of many other authors, too numerous to list. A bibliography appears at the ends of the major discussion chapters; I have leaned heavily on the authors of those works cited.

For permission to reprint copyrighted material I wish to thank the American Society of Heating, Refrigerating and Air Conditioning Engineers, the American Society of Mechanical Engineers, the National Fire Protection Association, the Carrier Corporation, the Mechanical Contractors Association, Rubber Manufacturer's Society, and Engineer's Press for the use of material from two of my earlier books; *Handbook of Florida Contractors* and *Journeyman General Mechanical Examination.* Also, for the use of material appearing in various catalogs, I wish to acknowledge the Allis-Chalmers Manufacturing Company, Ric-Wil Company, Mueller Brass Company, and the Marley Company.

I want to particularly thank Stephen Shelton for having read and criticized portions of the manuscript at an early stage , Sara Brockwell for her able assistance in preparing the text, and Cira Gonzalez for toiling through the typing and corrections of camera-ready copy.

John Gladstone

Naranja, Florida

CONTENTS

1. INTRODUCTION

This book is intended to fill a long existing need; preparing contractors and journeymen mechanics for their license examinations.

In recent years many states, cities and counties have begun to institute license examinations for refrigeration and air conditioning contractors where such examinations were not previously required. Building and zoning codes are being incorporated into law in the small, newly developed communities as well as in long established suburban municipalities that have previously functioned without such laws. Codes, as well as examinations for license, tend to become more strict as growth expands and as government becomes more complicated.

Although examinations are still not required in many communities, the wise contractor—and mechanic—will take his examination now, in the nearest community that offers one. There are several reasons why he should do this. Once licensed, the contractor need never again take an examination in that community and if, and when, the community expands or changes its incorporation status, the license expands with it. Frequently, municipalities will establish reciprocal agreements whereby a licensed contractor in one area may automatically operate in another. In some cases, a state which has never before had a licensing law, may write a law requiring all contractors in the state to take a license examination, but will extend the state license to a contractor who already holds such a license in any city of that state without an examination; this is one aspect of the "grandfather clause." Finally, every contractor should anticipate growth in his own business, he never knows when he will have to put in a job in an adjacent community.

For many years persons with long experience in the construction, electrical, plumbing, and air conditioning trades have been unable to get a license in their particular craft simply because they did not know how to take an examination. Many of these skilled tradesmen had not been in a classroom or examination room for over 20 years, and though they were skilled and capable out in the field, they were apprehensive and intimidated at the thought of sitting in a "classroom" and taking an examination. Most candidates for license examinations fail before they ever

1

set foot into the examination room.

"What is the examination like?" "Will there be lots
of 'trick' questions?" "What shall I study?" "Are there
many 'math' questions?" "How can I remember everything
in the code book?...there are thousands of pages." These
are some of the questions and fears that defeat the candidate as soon as he considers taking his exam. This
course is designed to give you the confidence and experience you need to take and pass, any contractor license
exam.

This book is not designed to teach you your trade or
profession...it is not intended as a high school diploma
or vocational training course...it is not intended as a
crib sheet to aid unqualified persons to get their license through some unethical scheme...it is not offered
as a substitute for studying. It is assumed that the
candidate has all the necessary qualifications, experience and education as set forth in the *Application For
Certificate Of Competency* and that he has complied with
all of the instructions and paid the required fee to the
Municipal, County, or State Board: in short, he is eli-
gible to take the exam.

HOW TO USE THIS STUDY COURSE

The course of study is carefully organized to develop your facilities for one purpose; passing the license examination. We might draw an analogy of a boxer training for his championship bout. He must conform to the discipline and organization imposed upon him by his manager and trainer, everything else in his life is pushed aside until after the big fight. Much of your success the day of the examination, depends on your ability to organize *now*. Do not skip through this course or hop about from one part to another. Follow the plan of organization as it is presented and understand each instruction before proceeding to the next.

You're a busy man--don't waste time. Make a study schedule and stick to it! The *STUDY SCHEDULE*, Figure 1.1, is designed to help you get organized. Fill in the maximum allowable study time you can afford. Investigation has proven that frequent short intervals of study at regular periods is more rewarding than less frequent but longer periods. However, this might not always be possible. Keep your schedule realistic. If your schedule is too ambitious and you can't meet it, it will break down your organization and you will lose the potential value of the course.

Assume you have marked your *STUDY SCHEDULE* to allow 1 hour of study each night plus 4 hours on Sunday. After the second week of study you discover that you have a conflict--owing to an association meeting or union meeting--on the second Tuesday of each month. It would then be wise to drop Tuesday out of the schedule completely...make that your night off each week and try to fit in other regular activities for the remaining Tuesdays, such as taking your wife to a movie. If your first plan does not work out well enough for you, change it to suit your particular circumstances, then stay rigidly with your new plan.

The examination announcement will usually list the reference material required for your particular exam. Certain books and pamphlets must be sent for by mail and are slow arriving. Do not procrastinate; get your necessary books and pamphlets as soon as possible. Remember, your profession requires certain reference material on your book shelf at all times. Money spent for this purpose is well spent, an investment you will never regret. Sometimes, one fact culled from a $25.00 book can save you hundreds, or even thousands of dollars on the job. Your accountant will advise you that expenditures for professional books are deductible from your income taxes.

3

THE 10 RULES FOR EFFICIENT STUDY

1. Make a schedule and stick to it. It will raise your level of personal efficiency. It will diminish emotional strain and lighten the burden. It will help you to master concentration. It will organize your entire family and reduce their interference with your program.

2. Study in the same quiet and well lighted place each time.

3. Keep the top of your desk or study table clear of all unrelated material. Do not wander off your course.

4. Start each study period by the clock, promptly, and end it the same way.

5. If your study sessions are long, take short rests periodically to relieve tension and stiffness.

6. Study is not reading: As you study, evaluate what you are trying to learn. Why is this expressed this way? What is it for? Can it be done another way?

7. Keep a pencil in your hand while you are studying, and a ruled pad or notebook alongside. A difficult or important passage should be written out; such an expression will help plant the thought firmly in your mind. Summarize ideas in the margin , underscore important, passages, rework hard to grasp ideas.

8. Keep a good dictionary of the English language on your desk. Be sure you know the meanings of all the words.

9. Don't get up from your work until it is time. If you need to have a smoke or nibble some pretzels with beer to keep relaxed during your study period, have these things ready *before* you start work.

10. Review your work constantly. Never, never, pass up a review because you feel you have already mastered that lesson.

S T U D Y   S C H E D U L E

This is your work plan; give it your meticulous attention. Develop a realistic schedule of the most possible hours you can devote to this program — then stick to it with an iron discipline! Draw an X through the hours you intend to set aside for study under each day and post this schedule in a visible place for constant reference.

	SUNDAY	MONDAY	TUESDAY	WEDNESDAY	THURSDAY	FRIDAY	SATURDAY
7:00							
8:00							
9:00							
10:00							
11:00							
12:00							
1:00							
2:00							
3:00							
4:00							
5:00							
6:00							
7:00							
8:00							
9:00							
10:00							
11:00							

FIGURE 1.1

PREPARING FOR THE EXAMINATION

An examination may be either "open book" or "closed book" or a combination of both. An "open book" exam means that the candidate is expected to solve problems and answer questions by referring to published books, other data, or notes. This is a pratical type exam and is usually based on given conditions which you are likely to encounter any time on an actual job. The "closed book" exam, or portion of exam, prohibits the use of any reference material in the test room. This is also, generally, a practical exam, but you may occasionally see some questions that are impractically presented. For example, you may be asked--on a refrigeration master exam--to give the operating pressure of Ref-717 at 290°F. Or, you may be asked--on an electrical master exam--to give the minimum required meter room dimensions for housing 23 meters. These examples of "closed book" memory type questions are seldom encountered; usually a memory type question will deal with a fundamental law or formula, something you work with every day. Do not attempt to memorize tables, charts or any questions and answers in this course. Although some of the questions you will see may reappear on your actual examination, it would be very unwise to attempt to memorize.

The "open book" section of the examination will really be a test of your ability to use available reference material, that is, you will be required to dig out certain information in a limited time. Time is the most critical thing about any examination, and the better organized you are to use your time properly, the better your chance to pass. Familiarize yourself thoroughly with the Table of Contents of each one of the volumes listed, and if possible, memorize them. If you have difficulty committing all of the contents to memory you should at least know the contents of the most important reference works. For example, you must remember that Pamphlets 90A, 90B, 91, 211, and 214 of the NFPA are all to be found in *Volume No. NFC-4*, and that "natural ventilation" will be found in the *ASHRAE Handbook of Fundamentals* while "industrial ventilation" will be found in the *ASHRAE Guide Volume Systems*. Study the contents of all of these books carefully. When you are thoroughly familiar with the contents, turn to the index and check the index against the text; this will give you an in-depth feel for each book.

Now, compile a "master table of contents" covering all of the material involved. This "master list of table of contents" will act as a subject guide index and could save the day for you. We have seen candidates taking an open book exam, flipping through page after page, in book after book, looking for the clue to a problem; these candidates always flunk.

Usually your weakness will show in a particular area. You may score high in psychrometrics and do poorly on control wiring. Or you may do very well on questions of a theoretical nature and show a weakness in building code regulations. This will be your clue on where to concentrate. No amount of memorization will help you...know your weakness and study hard in that area!

Whether an exam is "open book" or "closed book" an examination may be phrased in any of a number of styles. Often, the material for an exam, may be gathered by several different persons and comes from many sources. This will be reflected in the style of the questions; sometimes different styles are reflected in one examination owing to the different persons who were involved in writing the exam. Where an editor has been assigned to the job, the exam will be well organized and have an easy style.

Usually, questions will be of the *objective* type. That means you will not be required to do any design drawing or lengthy essay writing. Such questions are true and false, multiple choice, fill-in, etc. In other instances the exam—or portions of it—may require design work (usually of a limited nature), chart plotting, and/ or lengthy essay type answers. The present trend is towards objective type examinations answered on machine-scored cards. Most contractor, engineer, journeyman, and civil-service examinations around the country, are using the objective type examination with machine-scoring. In a few cases the largest part of the exam will be objective type supplemented with a small essay type section.

WHAT TO STUDY

Whether the examination is "open book" or "closed book" you must know what to study, and which books to use. Figure 1.2 shows a copy of the Florida State Licensing Board Notice to Applicants, dated January 14, 1974. It includes a list of *Suggested Reference Material* for each category in which an exam is given. No other books are allowed into the examination room in this case. In some cases the candidate may bring as many books as he feels he needs.

Figure 1.3 shows the *Notice and Instructions to Examinees* for a Dade County exam. Here the candidate is specifically told what he *must* bring and what he *may* bring. He is also told that psychrometric charts will be distributed by the testers, and that the candidates "will have some questions with direct reference to the Carrier Design Manual, part I --Load Estimating." In other cases no suggested reference list is published and the candidate must decide for himself what books to study from.

Usually, examination announcements will offer the candidate a list of recommended reference material. Such lists should receive your most careful attention and all books, pamphlets, codes and texts should be procured at the earliest possible time. Often, however, these lists are incomplete, incorrect, or obsolete. There is a tendency to reprint the same list year after year although some of the material has gone out of print or has been superseded by revised editions or new material.

Standard handbooks and reference manuals are usually updated periodically; in an era of fast-breaking new technologies, such new editions will completely obsolete the preceding ones. When a new standard is published, such as the *ASHRAE Guide and Data Book,* or the *National Electric Code,* only the new current edition is obtainable. The previous editions—having become obsolete—go out of print and are removed from available sources.

Writers of examinations as well as instructors for prep courses often neglect to update their material and are inclined to reuse old sources as a matter of convenience. National code revisions are usually adopted into examinations and college or vocational curriculum, long after the old code literature has gone into disuse. It is strongly recommended that examination candidates insist upon detailed instructions for obtaining reference material no longer available from the publisher. When a reference text has been declared out of print by a publisher, it should be removed from the reference list and automatically dropped as a study book for course work.

If a reference text is out of print and cannot be obtained, you should make a point of explaining this to your local examining board as soon as you have discovered it. If you do not receive any satisfaction from the board, you should then go beyond them and bring the matter to the attention of the county manager, or whoever is the higher authority. For too long now candidates have been obligated to take examinations for which no reference material has been available, simply because the exam writers were ignorant of the superseding edition or were using outdated material as a convenience.

STATE OF FLORIDA
DEPARTMENT OF PROFESSIONAL AND OCCUPATIONAL REGULATION
DIVISION OF OCCUPATIONS

FLORIDA CONSTRUCTION INDUSTRY
LICENSING BOARD
P.O. DRAWER 5257
TALLAHASSEE, FLORIDA 32301
AC 904 488-7010

REUBIN O'D. ASKEW
GOVERNOR

January 14, 1974

M E M O R A N D U M

TO: All Applicants submitting Applications for Examination.

FROM: The Florida Construction Industry Licensing Board
                    Allen R. Smith, Jr., Administrative Assistant

SUBJECT: Date of Examination

DUE TO THE FACT THAT ALL EXAMS WILL BE GIVEN AT THE SAME TIME,
YOU MAY ONLY APPLY FOR ONE CLASSIFICATION.

        Examination dates are as follows:

General
Builder
Residential - DIVISION A - March 22, 1974

Mechanical
Sheet Metal
Roofing
Air Conditioning
Pool -   DIVISION B - April 19, 1974

        The location of the examination is indicated on the
Admission Card enclosed with the application.
        The application MUST be POSTMARKED no later than
February 15, 1974 for both Division A and Division B examinations.

        The second examination for Fiscal year 1973-74 will be
administered June 14, 1974 for Division A and B. Applications
MUST be POSTMARKED no later than May 1, 1974 for Division A and
May 14, 1974 for Division B. The location of the exam is indicated
on the admission card enclosed with the application.

DIVISION A - General
             Builder
             Residential
DIVISION B - Mechanical   Air Conditioning
             Sheet Metal   Pool
             Roofing

                    FIGURE 1.2

Categories: A - Business Process Regulations and
                 Tax Laws
             B - Southern Standard Building Codes
             C - Safety Codes and Practices
             D - Technical/Material Knowledge

Look for your Classification under EACH Category listed.

SUGGESTED REFERENCE MATERIALS

CATEGORY A - (All Classifications)

Employers Tax Guide - Circular E (U.S. Treasury Department
1973).

Florida Contractors Licensing Law (Part II, Chapter 468,
Florida Statutes as amended July 1, 1973.)

Mechanics Lien Law (Part I, Chapter 713, Fla. Stat. 1971).

Publications of the Florida Department of Commerce:
(1) Florida Child Labor Law - "In a Nut Shell" 1971.
(2) The Florida Unemployment Compensation Law (Chapter 443,
     Fla. Stat. 1972.)
(3) Unemployment Compensation Code of Regulations (Chapter
     8-AU-1).
(4) Workmen's Compensation Law - 1972 (Chapter 440, Fla.
     Statutes as amended 1973.)

General and Supplementary Conditions of the Contract for
Construction (AIA Document A201 & A201/SC Federal Edition 1972.)

(Swimming Pool)- All of the above, in addition to:
Rules - State of Florida - Div. of Health. Chapter 100-S,
Swimming Pools & Bathing Places.

CATEGORY B

(General, Builder, Residential)-
Building Code Requirements for Reinforced Concrete (ACI
Standard 318-71).

Southern Standard Building Code - 1973 Edition

(Sheet Metal)-   NONE

(Air Conditioning)-
Southern Standard Mechanical Code - 1973

(Mechanical)-   Same as Air Conditioning

(Roofing)-   Southern Standard Building Code - 1973

(Swimming Pool)- Southern Standard Building Code - 1973

Building Code Requirements for Reinforced Concrete (ACI
Standard 318-71).

Southern Standard Mechanical Code - 1973

South Florida Building Code - 1973

FIGURE 1.2 (continued)

CATEGORY C

(General, Builder, Residential)-
Federal Department of Labor - Occupational Safety and
Health Administration - Safety and Health Regulations
for Construction. December, 1972 Vol. 37, No. 243 - Part
II.

(Sheet Metal)-   National Fire Code - Vol. 4 - 1973-74.
Building Construction and Facilities, particularly the
following:
(1)   No. 90A - Installation of Air Cond. & Ventilating
      Systems - 1972 - File 40-B-7.
(2)   No. 96 - Vapor Removal from Cooking Equipment - 1971.
(3)   No. 89-M Clearances for Heat Producing Appliances - 1971.
(4)   No. 90B - Installation of Warm Air Heating & Air
      Conditioning - 1965.

Recommended Construction & Shop Safety Practices for
the Sheet Metal Industry.Fla. Roofing & Sheet Metal Assn.

UL-181 Standard for Safety - Air Ducts.

Federal Department of Labor - Occupational Safety and
Health Administration - Safety and Health Regulations
for Construction. December, 1972 Vol. 37, No. 243 -
Part II.

(Air Conditioning)-
Same as Sheet Metal in addition to:
American Standard Safety Code for Mechanical Regrigeration.
ANCI - B9.1-1971.

(Mechanical)-   Same as Sheet Metal and Air Conditioning above, except UL-181
Standard for Safety - Air Ducts.
(Roofing)-   Federal Department of Labor - Occupational Safety and
Health Administration - Safety and Health Regulations
for Construction. December, 1972 Vol. 37, No. 243 -
Part II.

(Swimming Pool)-   Article No. 680 - National Electrical Code - 1971. Swimming
Pool Wiring - National Swimming Pool Institute. Washington,
D.C. - Spring, 1973.

Federal Department of Labor - Occupational Safety and Health
Administration - Safety and Health Regulations for Construction. December, 1972, Vol. 37, No. 243 - Part II.

CATEGORY D

(General Builder, Residential)-
Formwork for Concrete Structures (R. L. Puerifoy, Copyright
1964).

Standard Grading Rules for Southern Pine Lumber (1970)

FIGURE 1.2 (continued)

CATEGORY D - Con't
(General, Builder, Residential)

The Building Estimator's Reference Book - 18th Edition - Walker - 1973

Publications of the Portland Cement Assn:
(1) Design & Control of Concrete Mixtures(11th Edition)
(2) Concrete Masonry Handbook (1951)
(3) Cement Mason's Guide (1971) 2nd Printing 1973.

U.S. Standard Specification for Interior Lathing & Furring (ANSI A42.4,1967).

Gypsum Plastering (ANSI A42.1, 1964)

Gypsum Wall Board (ANSI A97.1, 1965)

Span Tables for Joist and Rafters (Abbr. Edition, Oct., 1971 - Revised 2/73).

Architectural and Building Trades Dictionary (by Burke, Dalzell and Townsend; 2nd Edition, 11th Printing, 1973, American Technical Society).

CRSI, Placing Reinforcing Bars - 2nd Edition 1972.

AISC - Manuel of Steel Construction

(Residential)-   Delete AISC - Manuel of Steel Construction
                 Delete CRSI - Placing Reinforcing Bars - 2nd Edition 1972
                 Delete Formwork for Concrete Structures(R.L.Puerifoy).
(Builder, Residential)-
                 All of the above in Category D, in addition to:

Publications of the National Forest Products Assn:
(1) Wood Construction Data - Numbers 1, 4, 6, & 7 (1970)
(2) National Designs Specifications - 1973 Edition & Supplement

(Sheet Metal)-   Sheet Metal and Air Conditioning Contractors Assoc. Publications as follows:
(1) AIA File 30-D-4 Manual for the Balancing and Adjustment of Air Distribution Systems. SMACNA (Revised 1973)
(2) Ducted Electric Heat Guide for Air Handling Systems. SMACNA #15 - 1st Edition 1971.

(3) Pressure Sensitive Tape Standards for Fibrous Glass Ducts - SMACNA #15 - 1st Edition 1973.

(Performance STD - AFTS - 100-73)
(Application STD - AFTS - 101-73)

(4) Air Handling Specifications. SMACNA #15 - 2nd Edition 1971.
(5) Residential Heating & Air Conditioning Systems - Minimum Installation Standards - SMACNA (NESCA) 1st Edition 1973.
(6) Fire Damper Guide for Air Handling Systems. SMACNA #15 - 1st Edition - 1970.

FIGURE 1.2 (continued)

CATEGORY D   Con't

(7) Duct Liner Applications Standards. SMACNA #15 - 1971.
(8) Fibrous Glass Duct Construction Standards. SMACNA #15 - 3rd Edition 1972.
(9) High Velocity Duct Construction Standards. SMACNA #15 - 2nd Edition 1969.
(10) Low Velocity Duct Construction Standards. SMACNA #15 - 4th Edition 1969.
(11) Architectural Sheet Metal Manual. SMACNA #7 - Thermal & Moisture Protection, Metal Roofing, etc.- 2nd Edition - June 1972.

Air Conditioning Cutter's Ready Reference by Morris - Revised 1971. (Business News Publishing Co., Birmingham, Michigan.)

(Air Conditioning)-Same as Sheet Metal, in addition to:

Air Conditioning Manual (Similar to Trane - 1965 Edition)

Handbook of Fundamentals - ASHRAE 1972.

Guide - Applications - ASHRAE 1971.

(Mechanical)- Same as Sheet Metal & Air Conditioning, in addition to:
Handbook of Fundamentals - ASHRAE 1972

Guide - Applications - ASHRAE 1971

Heating and Cooling Technical Manuel (Nat'l Assn. of Plumbing, Heating and Cooling Contractors).

Air Conditioning Manuel (Similar to Trane - 1965 Edition).

(Roofing)- Manuel of Build-Up Roof Systems by Griffin - 1970

Roofing Manual, National Roofing Contractor's Assn., 1st Edition 1972.

Publications of the Nat'l Forest Products Assn: (See list under Builder and Residential, Category D.

Grading Rules for Southern Pine Lumber (1970)

The Building Estimator's Reference Book (18th Edition - 1973 Walkers)

U. S. Standard Specifications for Interior Lathing and Furring (ANSI A42.4, 1967).

Gypsum Plastering (ANSI A42.1, 1964)

Gypsum Wall Board (ANSI A97.1, 1965).

Span Tables for Joist and Rafters (Abbrev. Ed., Oct., 1971, Rev. 2/73).

Architectural and Building Trades Dictionary. (By Burke, Dalzell, and Townsend; 2nd Edition, 11th Printing, 1973, American Technical Society.)

FIGURE 1.2 (continued)

METROPOLITAN DADE COUNTY · FLORIDA

ROOM 701
1351 N. W. 12TH STREET
MIAMI, FLORIDA, 33125
TEL: 377-6911

BUILDING AND ZONING DEPARTMENT

February 13, 1967

NOTICE & INSTRUCTIONS TO EXAMINEES

You are hereby notified that you have qualified to take the Mechanical
Masters Examinations and are to be present Saturday, February 18, 1967 at
8:00 A.M. in the Cafeteria of the Metro Justice Building, 1351 N.W. 12 St.
Miami, Florida.

All categories are open book, objective type examinations. You will not be
required to do any design drawing, etc. for problems. The questions will be
multiple choice, true & false, fill-in, etc.

A list of reference material has already been released. You are not limited
to this material, however, you may bring any reference books you wish;
each candidate is responsible for his own reference material. Exchange
of books and/or other materials between candidates will not be permitted.

Discussion, talking & exchanging of ideas between candidates will not
be permitted.

To the candidates taking categorial exams requiring psychrometric charts
for plotting the solutions to problems; these will be distributed with the
particular examinations. No other material will be distributed. Bring
your own pencils, slide-rules, pres/temp charts, straight edges, load
sheets for cooling, heating, refrigeration, etc.

All candidates taking categories Nos. 1, 1A, 2, will have some questions
with direct reference to the Carrier Design Manual Part 1 - Load Estimating.

The seating arrangement will be such as to separate all candidates in any
single category. Where more than one person from a single company or
organization is taking the examination, they too will be separately seated.

Should you have any questions regarding this notice prior to the examination,
phone the secretary to the Mechanical Contractors Examining Board at FR7-7101.

John Gladstone, Chairman
Examinations Committee
Mechanical Contractors
Examining Board.

JG/efb

FIGURE 1.3

Table 1.1 shows a list of Reference and Study Material compiled by the author. This is a basic study list and should be considered minimal for each particular category. If your examination announcement lists other specific books and pamphlets, then, of course, you are responsible for the additional information. Although few manufacturer's pamphlets appear in Table 1.1 (because they are too numerous to list in this limited space), they are often the most important source material.

Manufacturers and trade associations publish pamphlets and books containing excellent, reliable information concerning their products or areas of interest. These publications may cover design, safety, selection, application, installation and maintenance of products ranging from mosaic tile and plaster to boilers, pipe, and pollution control products. You should be thoroughly familiar with the available literature for your particular category. Locally, these may be available at your union hall, supply house, trade group, association, or manufacturer's representative. We suggest that you make every effort to acquire and study as much of this material as possible; it will provide answers to many of the practical questions appearing in your examination. You should also have a good dictionary of the English language.

In addition to listing obsolete references, the official announcement of recommended references may show incomplete or incorrectly spelled titles; or the publisher's or author's name may be misspelled or deleted. When you receive your official list from your municipality, check it against Table 1.1 for hints to correct titles.

CODES, STANDARDS AND STUDY MATERIAL

Bookstores do not generally carry the literature you are required to study from. If your local bookstore does carry some of the listed titles, good; if they do not carry them in stock, they may be willing to special order them for you. If your study material is not readily available through a local source, you will have to order by mail. In most cases your order will have to be accompanied by a check or money order; sometimes handling and postage charges are also required. In Table 1.2 you will find a list of the names and addresses of the most important associations and publishers of national codes and standards, as well as some of the publishers of other recommended reference texts and handbooks.

TABLE 1   SPECIALTY MECHANICAL CONTRACTOR

TITLE	Refrigeration and Air Conditioning Contractor	Air Conditioning Contractor (Unlimited)	Refrigeration Contractor (Unlimited)	Heating Contractor	Steam Generator Boiler and Boiler Piping	Warm Air Heating Contractor	Ammonia Refrigeration Contractor	Fire Sprinkler Contractor	Insulation Contractor	Mechanical Service and Maintenance	GENERAL MECHANICAL CONTRACTOR
ASHRAE Guide & Data Book: Applications (ASHRAE)	✓	✓	✓	✓	✓		✓				✓
ASHRAE Guide & Data Book: Equipment (ASHRAE)	✓	✓	✓	✓	✓		✓			✓	✓
ASHRAE Guide & Data Book: Systems (ASHRAE)	✓	✓	✓	✓	✓		✓				✓
Automatic Controls for Heat'g & Air Cond. (NESCA)						✓					✓
Ductulator (Trane Company)	✓	✓		✓		✓					✓
Fibrous Glass Duct. Construction Standards (SMACNA)	✓	✓				✓			✓		✓
Fire Damper Guide (SMACNA)	✓	✓		✓		✓				✓	✓
Gas Transmission of Piping Systems B31.8 (ASME)	✓										✓
HB of Air Cond't System Design: Carrier (McGraw)	✓	✓	✓	✓		✓	✓		✓	✓	✓
Handbook of Fundamentals (ASHRAE)	✓			✓						✓	✓
Heating Boilers, Section IV (ASME)				✓	✓						✓
High Velocity Duct Construction Standards (SMACNA)	✓	✓		✓		✓			✓	✓	✓
Low Velocity Duct Construction Standards (SMACNA)	✓	✓		✓		✓					✓
Manual for Balancing & Adjustment Air Dist. (SMACNA)	✓	✓		✓		✓				✓	✓
Manual J: Load Calculation (NESCA)	✓	✓		✓		✓					✓
Manual L: Installation Practice (NESCA)	✓	✓		✓		✓					✓
Mechanical Estimating Guide, 4/e: Gladstone (McGraw)	✓			✓		✓			✓	✓	✓
Modern Ref. & Air Cond.: Althouse, et al. (Goodheart)										✓	✓
National Fire Code, Volume 1 (NFPA)				✓		✓	✓				✓
National Fire Code, Volume 2 (NFPA)				✓		✓	✓	✓		✓	✓
National Fire Code, Volume 4 (NFPA)	✓	✓		✓		✓	✓	✓	✓		✓
National Fire Code, Volume 6 (NFPA)										✓	✓
Pipefitter's Handbook: Lindsey (Industrial Press)				✓	✓		✓			✓	✓
Power Piping B31.1.0 (ASME)	✓	✓	✓		✓		✓		✓	✓	✓
Ref., Air Cond. & Cold Storage: Gunther (Chilton)	✓	✓	✓				✓		✓	✓	✓
Refrigeration Piping B31.5 (ASME)	✓	✓	✓				✓				✓
Safety Code for Mechanical Ref. B9.1 (ASME)	✓	✓	✓				✓			✓	✓
Scheme for Identification Pipe Systems A13.1 (ASME)	✓	✓	✓	✓						✓	✓
Std. Boiler Operators Q & A: Elonka (McGraw)				✓	✓						✓
Thermal Insulation: Malloy (Van Nostrand/Reinhold)	✓	✓	✓						✓		✓
Trane Air Conditioning Manual (Trane Company)	✓	✓								✓	✓
Trane Refrigeration Manual (Trane Company)	✓	✓	✓				✓			✓	✓

Almost all municipal and county building codes in the United States are "referral codes" i.e., a code that adopts by referral, or conforms to a standard, already set by another code or national standard. The South Florida Building Code is such a document. For example: in the South Florida Building Code (SFBC) you may see:

> 4108.3 DIP TANKS: Dip tank operations shall conform to the standard "Dip Tanks" NFPA Pamphlet No. 34-1963 of the National Fire Protection Association, which is hereby adopted to supplement but not supersede the requirements set forth herein.

In the Southern Standard Building Code (SSBC) you may see:

> 901.4 MATERIAL (Sprinklers): Piping shall be as specified in "Standard of the National Fire Protection Association For the Installation of Sprinkler Systems--1963-1964".

Part 5 of the Refrigeration Code of the City of Los Angeles, read in part as follows:

> SEC. 95.12500 (b): All copper and brass refrigerant piping valves, fittings, and related parts used in the construction and installation of refrigerating systems shall conform to the American Standard Code for Pressure Piping, ASA B31. 1-1955.

There are many such referrals in the SFBC and SSBC. As a matter of fact, the SFBC refers to 38 separate standards in the Mechanical Sections alone. These references may be to ANSI, UL, NFPA, ASME, ASTM, etc., remember, it is your responsibility to know the law, and all of these Standards are law.

National Fire Codes, NFPA

Probably the most important and frequently quoted Standards are the NFPA pamphlets. These pamphlets are also the most difficult to locate because of their peculiar numbering system. However, a little familiarity will dispel all the difficulties. The NFPA publishes the National Fire Codes annually in 10 volumes. Each volume is divided into sections called "pamphlets," and each pamphlet is identified by a number. In the example given above for DIP TANKS the Standard was identified as NFPA Pamphlet No. 34-1963, this means that the SFBC adopted Pamphlet No. 34, as published in 1963. In the *List of Contents of the National Fire Codes*, Table 1.3, you will find a convenient index to all the pamphlets of the NFPA Code. Pamphlet No. 34 will be found in Volume I—Dip Tanks.

When a recommended reference list cites a designation such as "NFPA Pamphlet No. 34-1963," it is most practical to ignore the "1963" and get the latest edition of NFPA Volume 1. Obsolete Fire Code Pamphlets

TABLE 1.2

NAMES AND ADDRESSES OF IMPORTANT AGENCIES,
ASSOCIATIONS AND PUBLISHERS

Air Conditioning Refrigeration Institute
1815 North Fort Mayer Drive
Arlington, Va. 22209

American Gas Association
605 Third Avenue
New York, N.Y. 10016

American National Standards Institute Inc.
  (ANSI: Formerly American Standards
  Association, ASA and United States of
  America Standards Institute)
1430 Broadway
New York, N.Y. 10012

American Society of Heating, Refrigerating
  and Air Conditioning Engineers (ASHRAE)
345 East 47 Street
New York, N.Y. 10017

American Society of Mechanical Engineers
  (ASME)
345 East 47 Street
New York, N.Y. 10017

American Society of Sanitary Engineers
228 Standard Building
Cleveland, Ohio 44113

American Society for Testing & Materials
  (ASTM)
1916 Race Street
Philadelphia, Pa. 19103

American Water Works Association (AWWA)
2 Park Avenue
New York, N.Y. 10016

American Welding Society (AWS)
33 West 39 Street
New York, N.Y. 10017

Building Officials Conference of America
1313 East 60 Street
Chicago, Ill. 60637
(Basic Building Code)

Chemical Publishing Co., Inc.
212 Fifth Avenue
New York, N.Y. 10010

Committee on Industrial Ventilation (ACGIH)
P.O. Box 453
Lansing, Mich. 48902

General Services Administration (GSA)
Specifications Division
Building 197 Naval Weapons Plant
Washington, D.C. 20407

Industrial Press, Inc.
200 Madison Avenue
New York, N.Y. 10016

International Conference of Building
  Officials
50 South Los Robles
Pasadena, Calif. 91101
(Uniform Mechanical Code)

John Wiley & Sons, Inc.
605 Third Avenue
New York, N.Y. 10016

McGraw-Hill Book Company
330 West 42 Street
New York, N.Y. 10036

National Bureau of Standards
Products Standard Section
Office of Engineering Standards Service
Washington, D.C. 20234

National Environmental Systems Contractors
  Association (NESCA: Formerly National
  Warm Air Heating & Air Conditioning
  Association)
221 N. LaSalle Street
Chicago, Il. 60601

National Fire Protection Association (NFPA)
60 Batterymarch Street
Boston, Mass. 02110

Sheet Metal & Air Conditioning Contractor's
  National Association (SMACNA)
P.O. Box 3506
Washington, D.C. 20007

Southern Building Code Publishing Co.
116 Brown-Marx Building
Birmingham, Ala. 35203

Superintendent of Documents
U.S. Government Printing Office (GPO)
Washington, D.C. 20402

The Trane Company
La Crosse, Wis. 54601

Underwriters' Laboratories, Inc. (UL)
207 East Ohio Street
Chicago, Il. 60611

Van Nostrand-Reinhold Company
450 West 33 Street
New York, N.Y. 10001

CONTENTS OF THE NATIONAL FIRE CODE (NFPA) VOLUMES
AND PAMPHLETS: 1973-1974

　　　　In the following Table 1.3 the complete contents
of all ten volumes of the National Fire Codes are shown,
along with the date of the last revision for each particular pamphlet.　The NATIONAL FIRE CODES are revised
annually and published in late fall each year.　This
indispensable reference is composed of 206 Codes, Standards and Recommended References, contained in ten volumes.　Each Code or Standard is identified by number and
is referred to as a "pamphlet".　This document has been
reproduced with the permission of the National Fire
Protection Association.

　　　　Table 1.4 offers a Cross Index of the "Codes",
listing the pamphlets in numberical sequence for quickly
locating the volume number in which each pamphlet
appears.

TABLE 1.3

Vol. 1. Flammable Liquids, Ovens, Boiler-Furnaces

NFPA No. NFC-1

30	Flammable & Combustible Liquids Code
31	Oil Burning Equipment
32	Dry Cleaning Plants
321	Classification of Flammable Liquids
327	Cleaning Small Tanks
328	Manholes and Sewers, Flammable Liquids and Gases in
329	Underground Leakage of Flammable and Combustible Liquids
33	Spray Application Using Flammable Combustible Materials
34	Dip Tanks
35	Manufacture of Organic Coatings
36	Solvent Extraction
385	Tank Vehicles for Flammable & Combustible Liquids
386	Portable Shipping Tanks
393	Gasoline Blow Torches
395	Flammable Liquids on Farms, Storage of
85	Oil- and Gas-Fired Watertube Furnaces — One Burner
85B	Furnace Explosions in Natural Gas-Fired Multiple Burner Boiler-Furnaces
85D	Fuel Oil-Fired Multiple Burner Boiler-Furnaces
85E	Pulverized Coal-Fired Multiple Burner Boiler-Furnaces
86A	Ovens & Furnaces (Class A)
86B	Industrial Furnaces (Class B)
86C	Industrial Furnaces Using Special Processing Atmospheres (Class C)
704M	Fire Hazards of Materials, Identification System for

Vol. 2. Gases

NFPA No. NFC-2

37	Combustion Engines & Gas Turbines
50	Bulk Oxygen Systems
50A	Gaseous Hydrogen Systems
50B	Liquefied Hydrogen Systems at Consumer Sites
51	Welding & Cutting, Oxygen-Fuel Gas Systems for
51A	Acetylene Cylinder Charging Plants
51B	Cutting & Welding Processes
54	Gas Appliances and Gas Piping, Installation of
54A	Industrial Gas Piping & Equipment
56A	Inhalation Anesthetics
56B	Respiratory Therapy
56D	Hyperbaric Facilities
56E	Hypobaric Facilities
56F	Nonflammable Medical Gas Systems
56HM	Home Respiratory Therapy
57	Fumigation
58	Liquefied Petroleum Gases, Storage and Handling of
59	Liquefied Petroleum Gases at Utility Gas Plants
59A	Liquefied Natural Gas, Storage, Handling

Vol. 3. Combustible Solids, Dusts and Explosives

NFPA No. NFC-3

40	Cellulose Nitrate Motion Picture Film
41L	Model Rocketry Code
42	Pyroxylin Plastics in Factories
43	Pyroxylin Plastic Storage, Sale
43A	Liquid and Solid Oxidizing Materials
44A	Fireworks, Manufacturing, Transportation, Storage
48	Magnesium Storage, Handling
481	Titanium Storage, Handling
482M	Zirconium, Plants Producing
49	Hazardous Chemicals Data
490	Ammonium Nitrate Storage
492	Separation Distances of Ammonium Nitrate and Blasting Agents from Blasting Agents or Explosives
494L	Fireworks Law, Model State
495	Code for Explosive Materials
498	Explosives, Motor Vehicles, Terminals
60	Pulverized Fuel Systems
61A	Mfg. and Handling Starch
61B	Grain Elevators, Bulk Handling Facilities
61C	Feed Mills
61D	Agricultural Commodities for Human Consumption
62	Sugar & Cocoa, Dust Hazards
63	Industrial Plants, Dust Explosions
65	Aluminum Processing and Finishing
651	Aluminum Powder, Manufacture of
652	Magnesium Powder, Plants Handling
653	Coal Preparation Plants, Dust Hazards
654	Plastics Industry, Dust Hazards
655	Sulfur Fires, Explosions, Prevention
656	Spice Grinding Plants, Dust Hazards
657	Confectionery Manufacturing, Dust Hazards
66	Pneumatic Conveying Systems
664	Woodworking Plants, Dust Hazards
701	Flame-Resistant Textiles and Films, Fire Tests for
702	Flammability of Wearing Apparel

Vol. 4. Building Construction and Facilities

NFPA No. NFC-4

78	Lightning Protection Code
80	Fire Doors and Windows
80A	Protection from Exposure Fires
82	Incinerators, Rubbish Handling
89M	Clearances, Heat Producing Appliances
90A	Air Conditioning & Ventilating Systems
90B	Warm Air Heating & Air Conditioning Systems
91	Blower & Exhaust Systems
92M	Waterproofing, Drainage of Floors
96	Removal of Smoke & Grease-Laden Vapors from Commercial Cooking Equipment
97M	Glossary of Heating Terms
101	Life Safety Code
102	Tents, Grandstands & Air-Supported Structures Used for Places of Assembly
203M	Roof Coverings
204	Smoke & Heat Venting Guide
206M	Building Areas & Heights
211	Chimneys, Fireplaces & Vents
214	Water Cooling Towers
220	Building Types, Standard
241	Building Construction and Demolition Operations
251	Fire Tests, Building Construction & Materials
252	Fire Tests, Door Assemblies
255	Building Materials, Tests of Surface Burning Characteristics of
256	Fire Tests, Roof Coverings
257	Fire Test of Window Assemblies
703	Fire Retardant Treatments, Building Materials

Vol. 5. Electrical

NFPA No. NFC-5

70	National Electrical Code
70A	Electrical Code for One- and Two-Family Dwellings
70L	Model State Electrical Law
75	Electronic Computer Data Processing Equipment, Protection of
76A	Elec. Sys. for Health Care Facilities
76CM	High-Frequency Electrical Equipment in Hospitals
79	Electrical Metalworking Machine Tools
493	Intrinsically Safe Process Control Equipment
496	Purged and Pressurized Enclosures for Electrical Equipment

Vol. 6. Sprinklers, Fire Pumps & Water Tanks

NFPA No. NFC-6

13	Sprinkler Systems, Installation
13A	Sprinkler Systems, Maintenance
13E	Fire Department Operations in Properties Protected by Sprinkler, Standpipe Systems
16	Foam-Water Sprinkler & Spray Systems
20	Centrifugal Fire Pumps
21	Steam Fire Pumps, Maintenance
22	Water Tanks
24	Outside Protection

TABLE 1.3 (continued)

Vol. 7. Alarm & Special Extinguishing Systems

NFPA No. NFC-7

11	Foam Extinguishing Systems
11A	High Expansion Systems
11B	Synthetic Foam and Combined Agent Systems
12	Carbon Dioxide Extinguishing Systems
12A	Halon 1301 Extinguishing Systems
12B	Halon 1211 Extinguishing Systems
14	Standpipes & Hose Systems
15	Water Spray Fixed Systems
17	Dry Chemical Systems
18	Wetting Agents
26	Supervision of Valves
291	Fire Hydrants, Uniform Markings
292M	Water Charges, Private Protection
69	Explosion Prevention Systems
71	Central Station Signaling Systems
72A	Local Protective Signaling Systems
72B	Auxiliary Signaling Systems
72C	Remote Station Signaling Systems
72D	Proprietary Signaling Systems
73	Public Fire Service Communications
74	Household Fire Warning Equipment

Vol. 8. Portable & Manual Fire Control Equipment

NFPA No. NFC-8

2M	Model Drafts for Enabling Legislation
3M	Hospital Emergency Preparedness
4	Organization for Fire Services
4A	Fire Department, Organization of a
6	Industrial Fire Loss Prevention
7	Fire Emergencies, Management
8	Effects of Fire on Operations, Management Responsibility for
9	Training Reports and Records
10	Portable Fire Extinguishers, Installation
10A	Portable Fire Extinguishers, Maintenance & Use
10L	Model Enabling Act, Portable Fire Extinguishers
182M	Vaporizing Liquid Agents, Hazards
19	Automotive Fire Apparatus
19B	Respiratory Protective Equipment for Fire fighters
191	Portable Pumping Units
193	Fire Department Ladders
194	Fire Hose Couplings, Screw Threads
196	Fire Hose
197	Initial Fire Attack, Training Standards on
198	Fire Hose, Care of
25	Water Supply Systems for Rural Fire Protection
27	Private Fire Brigades
295	Wildfire Control
601	Watchman or Guard Manual
601A	Standard for Guard Operations in Fire Loss Prevention
604	Salvaging Operations
901	Uniform Coding for Fire Protection
901AM	Field Incident Manual

Vol. 9. Occupancy Standards & Process Hazards

NFPA No. NFC-9

46	Lumber, Outdoor Storage
46A	Wood Chips, Outdoor Storage
46B	Outdoor Storage of Logs
47	Lumberyards, Retail, Wholesale
56C	Hospital Laboratories
68	Explosion Venting, Guide
77	Static Electricity, Recommended Practices on
81	Fur Storage, Cleaning
87	Piers and Wharves
88A	Parking Structures
88B	Repair Garages
224	Homes, Camps in Forest Areas
231	General Storage, Indoor
231A	General Storage, Outdoor
231B	Storage of Cellular Rubber and Plastics
231C	Rack Storage of Materials
232	Record Protection
232AM	Archives and Record Centers
501A	Mobile Home Parks
501B	Mobile Homes
501C	Recreational Vehicles
501D	Recreational Vehicle Parks
513	Motor Freight Terminals
602	Community Dumps
801	Facilities Handling Radioactive Materials
802	Nuclear Reactors
910	Protection of Library Collections
911	Protection of Museum Collections

Vol. 10. Transportation

NFPA No. NFC-10

302	Motor Craft
303	Marinas & Boatyards
306	Control of Gas Hazards on Vessels
307	Marine Terminals, Operation
311	Ship Fire Signal
312	Vessels During Construction
402	Aircraft Rescue, Fire Fighting Standard Operating Procedures
403	Aircraft Rescue, Fire Fighting Services at Airports
406M	Fire Dept. Handling Crash Fires
407	Aircraft Fueling
408	Aircraft Fire Extinguishers
409	Aircraft Hangars
410A	Aircraft Electrical Maintenance
410B	Aircraft Oxygen Maintenance
410C	Aircraft Fuel System Maintenance
410D	Aircraft Cleaning, Painting & Paint Removal
410E	Aircraft Welding Operations in Hangars
410F	Aircraft Cabin Cleaning Operations
412	Evaluating Foam Fire Fighting Equipment
414	Aircraft Rescue, Fire Fighting Vehicles
415	Aircraft Fueling Ramp Drainage
416	Airport Terminal Buildings
417	Aircraft Loading Walkways
418	Roof-top Heliport Construction and Protection
419	Airport Water Supply Systems
421	Aircraft Interior Fire Protection
422M	Aircraft Fire Investigators Manual
505	Industrial Trucks, Powered
512	Truck Fire Protection

TABLE 1.4

CROSS INDEX FOR LOCATING NFPA PAMPHLETS 1973-74

Pamphlet Number	Volume Number	Pamphlet Number	Volume Number	Pamphlet Number	Volume Number
2 M	8	33	1	61 B	3
3 M	8	34	1	61 C	3
4	8	35	1	61 D	3
4 A	8	36	1	62	3
6	8	37	2	63	3
7	8	40	3	65	3
8	8	41 L	3	66	3
9	8	42	3	68	9
10	8	43	3	69	7
10 A	8	43 A	3	70	5
10 L	8	44 A	3	70 A	5
11	7	46	9	71	7
11 A	7	46 A	9	72 A	7
11 B	7	46 B	9	72 B	7
12	7	47	9	72 C	7
12 A	7	48	3	72 D	7
12 B	7	49	3	73	7
13	6	50	2	74	7
13 A	6	50 A	2	75	5
13 E	6	50 B	2	76 A	5
14	7	51	2	76 CM	5
15	7	51 A	2	77	9
16	6	51 B	2	78	4
17	7	54	2	79	5
18	7	54 A	2	80	4
19	8	56 A	2	80 A	4
19 B	8	56 B	2	81	9
20	6	56 C	9	82	4
21	6	56 D	2	85	1
22	6	56 E	2	85 B	1
24	6	56 F	2	85 D	1
25	8	57	2	86 A	1
26	7	58	2	86 B	1
27	8	59	2	86 C	1
30	1	59 A	2	87	9
31	1	60	3	88 A	9
32	1	61 A	3	88 B	9

TABLE 1.4 (Continued)

Pamphlet Number	Volume Number	Pamphlet Number	Volume Number	Pamphlet Number	Volume Number
89 M	4	295	8	482 M	3
90 A	4	302	10	490	3
90 B	4	303	10	492	3
91	4	306	10	493	5
92 M	4	307	10	494 L	3
96	4	311	10	495	3
97 M	4	312	10	496	5
101	4	321	1	498	3
102	4	327	1	501 A	9
182 M	8	328	1	501 B	9
191	8	329	1	501 C	9
193	8	385	1	501 D	9
194	8	386	1	505	10
196	8	393	1	512	10
197	8	395	1	513	9
198	8	402	10	601	8
203 M	4	403	10	601 A	8
204	4	406 M	10	602	9
206 M	4	407	10	604	8
211	4	408	10	651	3
214	4	409	10	652	3
220	4	410 A	10	653	3
224	9	410 B	10	654	3
231	9	410 C	10	655	3
231 A	9	410 D	10	656	3
231 B	9	410 E	10	657	3
231 C	9	410 F	10	664	3
232	9	412	10	701	3
232 AM	9	414	10	702	3
241	4	415	10	703	4
251	4	416	10	704 M	1
252	4	417	10	801	9
255	4	418	10	802	9
256	4	419	10	901	8
257	4	421	10	901 AM	8
291	7	422 M	10	910	9
292 M	7	481	3	911	9

are harder to find than last year's newspaper, and local
sources rarely stock any pamphlets at all. Technical
book specialists may stock all ten volumes of the
National Fire Code in current edition; they do not stock
individual pamphlets.

American National Standards Institute; ANSI

Both the Southern Standard Building Code and the
South Florida Building Code, contain many references to
ANSI Standards, and all contractors should be thoroughly
familiar with these references; for examinees, it is
imperative. The recommended reference list for exams
usually list a number of ANSI Standards. An ANSI Standard implies a consensus of those substantially concerned with its scope and provisions. There are over
3000 ANSI Standards. In many cases another organization
such as ASME or ASTM may be the sponsor of a particular
Standard and carry another designation number. For
example; the NFPA 31-1965 and the ANSI Z95.1-1965 are
identical and may be referred to by either the NFPA or
ANSI designation number: the American Concrete Institute, "Building Code Requirements for Reinforced Concrete" ACI 318-63, is identical to ANSI A89.1-1964.
Most frequently, references are made to the designation
number of the sponsoring organization; NFPA 31-1965 and
ACI318-63, rather than the ANSI numbers.

Code for Pressure Piping

The Code for Pressure Piping is sponsored by the
American Society of Mechanical Engineers and consists
of 8 Sections. Of these, 7 have been published and 1
is in preparation and has not yet been published. The
Sections are as follows:

1 Power Piping Systems(1967)	B31.1.0	$10.50
2 Fuel Gas Piping Systems(1968	B31.2	$ 4.00
3 Petroleum Refinery Piping(1966)	B31.3	$ 6.00
4 Liquid Petroleum Transportation Piping(1971)	B31.4	$ 6.25
5 Refrigeration Piping Systems (1966)	B31.5	$ 5.00 includes 1968 addenda.
6 Chemical Process Piping (in preparation)	B31.6	
7 Nuclear Power Piping(1969)	B31.7	$16.25 includes 1971 addenda.
8 Gas Transmission & Distribution Piping(1968)	B31.8	$ 8.00 includes 1971 addenda.

Each Section is headed by a committee whose function it is to keep abreast of the current technological improvements in new materials, fabrication practices, and testing techniques; and to keep the code updated to permit the use of acceptable new developments. The Committee issues periodic revisions as required and ANSI approves new editions at intervals of 3 to 4 years.

Boilers and Pressure Vessel Code: ASME

The ASME Boiler and Pressure Vessel Code is the work of the American Society of Mechanical Engineers; Boiler and Pressure Vessel Committee. This code is independent of ANSI and consequently is not referenced by the ANSI numbering system. It consists of the following Sections which are shown with ASME's list price—for the covenience of contractor examination candidates:

Section	Title	Price
1	Power Boilers	$17.00
2	Material Specifications	
	Part A Ferrous	$27.00
	Part B Nonferrous	$22.00
	Part C Welding Rods	$12.00
3	Nuclear Power Plant Components	$35.00
4	Heating Boilers	$16.00
5	Nondestructive Examination	$15.00
6	Care and Operation of Heating Boilers	$ 8.00
7	Care of Power Boilers	$ 8.00
8	Pressure Vessels	
	Division 1	$22.00
	Division 2: Alternative Rules	$22.00
9	Welding Qualifications	$11.00
10	Figerglass Reinforced Plastic Pressure Vessels	$11.00
11	Inservice Inspection of Nuclear Reactor Coolant Systems	$ 6.00
	Case Interpretations Book	$20.00

It is a simple expedient for the SFBC to adopt the entire Boiler and Pressure Vessel Code as a standard for Chapter 40, but it is an inconsideration to hold an examination candidate responsible for the entire "Code". Examination references, as well as municipal and state building codes, should cite the Boiler and Pressure Vessel Code by Section numbers as shown above. Those sections not applicable should be deleted from the recommended reference. Aside from the cost of $237.00, it is no small matter to study the entire "Code."

Candidates for all contractor categories, and particularly the mechanical categories, should carefully
examine the titles and contents of the "Standards" in
the recommended reference lists before embarking on a
study program. The LIST OF REFERENCES AND STUDY MATE-
RIAL in Table 1.1 has been carefully checked and veri-
fied for applicability and availability: if any of
these appear on the reference list circulated by the
examining board, you should not hesitate to incorpo-
rate them into your study program. Table 1.5 gives a
cross index of the standards applying to the mechanical
section of the South Florida Building Code; it is typical of the standards adopted by most of the codes
around the country.

U.S. Government Standards: GPO

Many Standards are published by the U.S. Government. These include the National Bureau of Standards,
FHA, HUD, PHS, etc. With few exceptions, they are all
available from the Government Printing Office, Superintendent of Documents, Washington, D.C. 20402. To
avoid long delay, it is suggested that you send your
order directly to the GPO with check or money order in
the amount listed. The Government will not ship any
of these documents without payment in advance, nor will
they process the shipment until your check has cleared
the bank. A postal money order could speed matters,
but do not send postage stamps.

Table 1.6 lists some important U.S. Government
publications along with list prices for your convenience in ordering. National Bureau of Standards publications are also available from the GPO (with some
exceptions), for the complete list of NBS publications
and availability, write to the National Bureau of
Standards, Office of Engineering Standards Services,
Washington, D.C. 20234.

SFBC & SSBC Versus National Standards

Occasionally a conflict will be found between a
Municipal Code and a national Standard. When this
occurs remember that your local code takes precedence
over any other Standard, even a self-adopted one. For
example, the SFBC, 4802 (c), page 443, calls for a
"complete change of air every three minutes" for dry-
cleaning plants. The NFPA Volume 1 (66-67) page 32-9,

TABLE 1.5

CROSS INDEX OF THE MAJOR STANDARDS PERTAINING TO THE
MECHANICAL SECTIONS OF THE SOUTH FLORIDA BUILDING CODE (1970)

Column 1 gives the adopted standard, column 2 gives the construction covered, column 3 gives the paragraph where the SFBC reference appears, column 4 gives (where applicable) the NFPA volume number in which each particular NFPA pamphlet appears.

STANDARD	CONSTRUCTION COVERED	SFBC SECTION	NFPA VOLUME
ANSI A 17.1	Elevators and escalators	3201.2	-
NFPA 90 A	Fire resistance protection	3703.6	4
NFPA 13	Sprinkler systems	3801.2	6
NFPA 14	Standpipes	3803.1	7
NFPA 22	Pressure tanks	3804.4	6
NFPA 54	Vent connections for gas appliance	3905.2	2
NFPA 90 B	Smoke Pipe clearances	3905.3	4
NFPA 89 M	Clearances for heat producing appliances	4001.4	4
NFPA 31	Oil burning equipment	4004	1
ASME	"Boiler & Pressure Vessel Code" boilers	4006	-
NFPA 30	Flammable and combustible liquids	4102	1
NFPA 31	Oil burning equipment	4102	1
NFPA 90 A	Nonresidential air conditioning	4103	4
NFPA 90 B	Residential warm air and air conditioning	4103	4
NFPA 96	Ventilation & restaurant cooking equipment	4103	4
NFPA 91	Blowers and exhausts	4103	4
NFPA 664	Woodworking plant dust exhaust	4103.5	3
NFPA 33	Spray finishing: paint booths	4107.2	1
NFPA 34	Dip tanks	4107.3	1

TABLE 1.5 (Continued)

STANDARD	CONSTRUCTION COVERED	SFBC SECTION	NFPA VOLUME
C 272	PVC Schedule 40 condensate drains	4606.7	-
Fla. Health Dept.	Closed well systems	4616.4	-
NFPA 54	Installation of gas appliances & piping	4702.1	2
NFPA 58	Storage of LP gas	4702.2	2
NFPA 33			1
NFPA 90 A			4
NFPA 90 B	Forced ventilation: outside air	4801.2	4
NFPA 91	"Standards of good practice"		4
NFPA 96			-
ASHRAE Guides			
ANSI B 9.1			-
ANSI B 31.1			-
ANSI B 31.2	Air conditioning and refrigeration	4902	4
NFPA 90 B			4
NFPA 214			4
NFPA 90 A			-
ASHRAE Guides	"Standards of good practice"	4902	-

TABLE 1.6

IMPORTANT REFERENCES FROM THE U.S. GOVERNMENT PRINTING OFFICE

PUBLICATION	PRICE
Minimum Property Standards for Multifamily Housing, FHA No. 2600 (Includes Mechanical Ventilation and Air Conditioning), November 1963	$2.50
Minimum Property Standards for One and Two Living Units, FHA No. 300, May 1963	2.00
Minimum Property Standards for Low Cost Housing FHA No. 18, June 1963	25 cents
Minimum Property Standards for Urban Renewal Rehabilitation, FHA No. 950	40 cents
Minimum Property Standards for Swimming Pools, HUD No. PG-30 (Formerly FHA No. 550) June 1967	25 cents
Manual of Septic Tank Practice, PHS No. 526, 1967	35 cents

calls for an "air change every 6 minutes". The SFBC
supersedes the NFPA

WHAT TO BRING TO THE EXAM ROOM

In most cases the official Instructions to Examinees
will list the items you may bring with you, or "must
furnish", in addition to the standard reference books.
If you have not been instructed what to bring, you should
call the Contractor's Section of the Building & Zoning
Department or the Licensing Board having jurisdiction,
and ask them. In some cases you will be limited to a
pencil—or an electrolizer pencil, for marking machine-
scored answer cards. Or there may be no limit, in
which case you may even bring an adding machine or calculator.

By all means bring your reading glasses if you require them; it would be a good idea to bring a spare
pair...just in case. Also bring 2 pencils, preferably
electrolizer type for electric machine scoring, and a
small pocket knife or pocket pencil sharpener. A pocket scale rule and a pocket slide rule might come in
handy too.

If your exam is open book and your list is unlimited, try to be selective about what you bring; too
many books will only be cumbersome and clutter your
work table or desk. Table 1.1 will help you select
the basic reference texts. In addition you may have
some manufacturer's catalog from which you have been
accustomed to work, these will be helpful because you
are familiar with them. The key to any open book exam
is knowing where to find the information: familiarize
yourself thoroughly with all recommended reference
works—know your code books. If you must memorize anything, memorize the Table of Contents and Index of your
most important reference books!

THE DAY OF THE EXAMINATION

Almost as important as the examination day, is the
day before. Try to schedule an easy, relaxed day before the exam. Do not do any unnecessary work on exam
night; relax and get a good night's sleep. Don't try
to cram a lot of studying in on the last night; if you
do not know your stuff now, it's too late.

Allow yourself plenty of time to arrive at the
examination room 15 minutes ahead of start time. This
is important advice. Hazardous driving, hurrying,

tension of any kind...will affect your thinking and recall faculties. Arrive early—keep confident—keep relaxed.

Concentrate on the examination. Read the directions carefully, then read them again. If there are verbal instructions, listen attentively. Maintain your sense of organization: get an overview of the exam before you start work and try to judge how much time you can spend on each section.

Keep your work neat. Do not sacrifice a properly marked paper in the interest of speed. Mark each answer carefully and accurately in the proper space provided, especially on machine-scored cards. Mechanical scoring devices have no concept of your good intentions; many candidates have failed examinations because of improperly marked papers.

Examinations are often weighted, i.e., a greater value is placed on one part of the examination than another. If this is the case it will be so stated in your instructions. Try to pace yourself accordingly and spend an equivalent amount of time to the given weight or value. Do not spend half of your allowable time on a section of the exam that is weighted at 25%.

Examinations are seldom graded on a penalty system, therefore you should answer every questions; you are not penalized for an omission. If you do not know an answer, guess at it and take your chances...a chance is better than no answer.

If you find a section of the examination particularly puzzling, move on to the next section, leaving the difficult one to return to later. After you have completed all the easy sections, go back and clean up the tough ones.

Use all of the time allotted to you. If the time limit for the exam is 4 hours, do not leave the room in 3 hours. Spend all of the remaining time reviewing your answers: if you find one error, it was worth the extra hour. But remember, do not plan to do a total reevaluation of your answers. Just review...make certain you marked every question, wrote each answer in the proper place, said what you meant in each case, and clearly understood each question.

If there is a time limit for each section (examinations are sometimes organized in this manner) do not spend more than your allowable time for any section finished or not, move on to the next section.

If you feel stress or tension, take a break...right now. Do not postpone a break when you feel too tense to work. A brief break for a minute or two...a trip to the water fountain or men's room...may mean the difference in passing. If the rules prohibit your leaving your seat, try closing your eyes and taking some long stretches in varying positions, pumping your muscles, rotating your head, etc.

TACTICS USED BY TEST WRITERS
AND THE STRATEGY REQUIRED TO BEAT THEM

The scope, contents and style of an examination depends, of course, upon the writer or writers. All examinations are of limited scope i.e., an examination to text your knowledge or refrigeration is limited to refrigeration questions. There are only so many questions that can be composed on the subject without duplication. Variations on a theme may be diverse but the theme remains basic. Where problem solving is required as part of the exam, such as, solving the refrigeration load for a given set of conditions; the given conditions may vary greatly, but the basic steps for finding the solution remain unchanged.

Your examination may be written by a professional examination group, a college professor, a government official, a group of contractors acting as a "Board", or a registered engineer. Who writes the examination will determine the contents and style of questions; it is often possible to tell who wrote the exam by the way it reads. Government officials are inclined to load the examination with questions related to code, a contractor board with "pratical" type questions, and a college professor with "theoretical" type questions.

To avoid the possibility of fraud, or defraud, identical examinations are seldom, if ever, given twice. In those cases where an examination is given more than once, it alternates with other exams so that the same examination is never given in succession and will only reappear after 3 or 4 years. Usually the examiners will maintain a bank of questions for each category; the bank may contain several hundred or even thousand questions. When the examination is being prepared (say 50 questions are required for a particular exam) the questions are pulled from the bank at random. The chances are that some of the questions from an earlier examination will always reappear.

It is not unusual for different municipalities to obtain questions from a similar body, and in some cases, municipalities exchange examinations and exam information. Regardless of source or intent, all exams of a particular category will have some similar, if not identical questions. This is true because of the universality and—as mentioned above—the limited scope of any subject. You might be asked:

One ton of refrigeration is equal to 12,000 Btu/hr. True, or false: or;

$$I = \frac{W}{E} \quad \text{True, or false?}$$

These are universal truths, and both of these questions will appear, as shown , or in some variation, in any examination in the country. The chances are that some of the questions—or variations of them—that appear in the SAMPLE EXAM, will appear on your actual examination.

WHEN YOU GET THE SIGNAL TO BEGIN

RELAX: Read the directions carefully and be certain that you know whât to do...and what not to do. You may be instructed not to write on your test booklet or question paper, but only on your answer sheet. You may be instructed not to write your name on any sheet, but only the number given you with your test. Follow instructions carefully; any violation could disqualify you.

Having read the directions and identified yourself in the proper spaces, put down your pencil and scan the entire exam. This overview of the whole examination is an important part of your strategy; it gives you perspective. The perspective overview introduces you to the style of the exam, it familiarizes you with the general tone and cadence of the exam. You may detect several styles in different sections, indicating different writers. You will have a "feel" that it's an easy or tough exam, that you are well prepared through study, or that many questions on control wiring (your weakness) appear. Further, a good overview often affords a clue to early questions in the exam through some revelations which appear further along. And, knowing how many questions there are will give you a fair indication of how much time you can spend per question..now you can pace yourself!

THE ESSAY QUESTION

   An essay question requires you to compose an answer
and write it out. Most modern examinations do not use
this type of question because it can not be answered for
machine-scoring; it must be read by the examiner and
graded accordingly.

   An essay type question requires more than a knowledge of your subject—you need to be able to express
yourself, or explain that knowledge to others. Before
answering, make certain that you are expressing your
thoughts correctly. Examine the phrase in your mind
before you set it down on paper; there may be a better
way to express your thought, or perhaps the word position should be changed. For example, you write:

   "The man ran wildly down the street after the dog
   in red pajamas".

you meant:

   "The man in red pajamas ran wildly down the street
   after the dog".

or:

   "He is the man in the blue car's brother".

you meant:

   "He is the brother of the man in the blue car".

   Watch out for the "multiple answer trap". An
essay question may have more than one answer. Write
all the answers which may apply; the examiner could be
looking for the one you omitted. For example:

   Q. One or more units of a system are not heating.
       What is the probable cause?

   Think of *all* the likely causes and include them in
your answer:

   A. The air eliminators are not opening...steam
       trap is stuck...dirt in the lines...inoperative
       valve...failure in the control system...pocket
       in the steam line.

Another example:

Q. Why is the suction riser in a refrigeration system usually of smaller diameter than the horizontal lines?

Try to picture the circuit in your mind and answer the question fully: half an answer will be marked wrong.

A. To increase the velocity of the gas and insure the return of oil to the compressor.

Either half of the answer is correct, but the examiner would probably consider it incomplete...don't chance it.

THE COMPLETION, or FILL-IN QUESTION

This is a suggestive type question; some persons are more responsive to this type of question than others. The answer usually comes to mind quickly. If it doesn't, don't fret long; move on to the next questions and come back to that one later. A fresh look at it for the second time may suggest the answer. Example:

Q. The secondary high voltage windings of a small power transformer are color coded _____

You would be expected to write in the word "red" into the blank space. If you do not know color coding, do not waste any time with that question; your experience will tell you that it is either black, green, white or red...take your guess and move along.

The fill-in question is not frequently used in modern exams, but you may encounter it.

THE MULTIPLE CHOICE QUESTION

This is the most popular type question, the kind you are most likely to encounter on your next examination. Multiple choice questions are written for most civil service exams, college entrance exams, municipal, county and state board exams, vocational training courses, and regular college examinations.

Although "trick" or "catch" questions will rarely appear on your exam, the questions may be framed in such a way as to throw you off guard. You are offered a cluster of either 4 or 5 answers to each question and you must select only one answer. Very often all the an-

swers seem possible; you must select the most correct
answer. When a skilled examiner has written the examination, the multiple choice question can be a real brain
wringer. Do not hurry...examine each question carefully,
regardless of how simple the answer looks at first
glance...the obvious looking answer is often placed
there just to force you to think through all the possible answers.

Attacking the Multiple Choice Question:

1. Look for the qualifying word in the question
   e.g.
   always...never least...most often...rarely
   most likely...least likely sometimes...always
   smallest...largest probably...possibly
   maximum...minimum best, most advisable,
   usually, greatest, chiefly, etc.

Any of these qualifiers offer a clue to the correct
answer.

2. Don't assume the meaning of a word. The use of
   the abbreviation, e.g., above, was deliberately
   used to check your comprehension. If you do not
   know what e.g. means look it up in your dictionary. You should find it in the back section,
   "Abbreviations". If your dictionary does not
   carry such a list of abbreviations, it is un-
   suitable...get a good one. The meaning of
   words is critical to any exam; if you are un-
   familiar with any words used in this study
   course or Sample Test, use your dictionary.
   As part of your study curriculum you should
   have a good *Glossary of Trade Terms*.

3. If you are unsure of the answer, apply the
   principle of elimination. That is, first
   eliminate the choices that you know are wrong;
   this narrows your choice down to 3 or perhaps
   2 possible answers. Sometimes, the process
   of elimination will lead you directly into the
   correct choice.

4. Never allow yourself to be influenced by an
   answer "pattern". There is no pattern of
   answers. When answers appear to be following
   a certain pattern, it is usually coincidental.
   If the questions are deliberately placed in
   pattern...watch out! You may get a series of

questions like:

Q. Bare wire ends are spliced by the

    a. Western Union method
    b. rat-tail joint method
    c. fixture joint method
    d. all of the above

... in 3 or 4 successive questions the answer will be "d", as in the above. Then you will get:

Q. How are solderless connectors installed on conductors?

    a. crimped on
    b. lugged on
    c. bolted on
    d. all of the above

... the "d" pattern is broken, when you least suspect it!

In the following example you see an actual question which has appeared on many exams and is usually answered wrong. Most candidates would know the correct answer if they would think about it for a moment, but the question is phrased in such a way as to appear simple:

Q. A rotating vane anemometer is an indicating instrument that registers an air current in

    a. mph
    b. cfm
    c. fpm
    d. feet

Over ninety percent of examinees will mark "c"; the correct answer is "d". Do not rush into an answer, think for a moment before you make your selection.

THE MACHINE SCORED ANSWER SHEET

Figure 1.4 is a reproduction of an actual Test Answer Sheet frequently used for contractor examinations. Where this system is used, you will be given a Test Answer Sheet along with your examination papers. Each question will be numbered and you will be required to mark the corresponding number on the Answer Sheet.

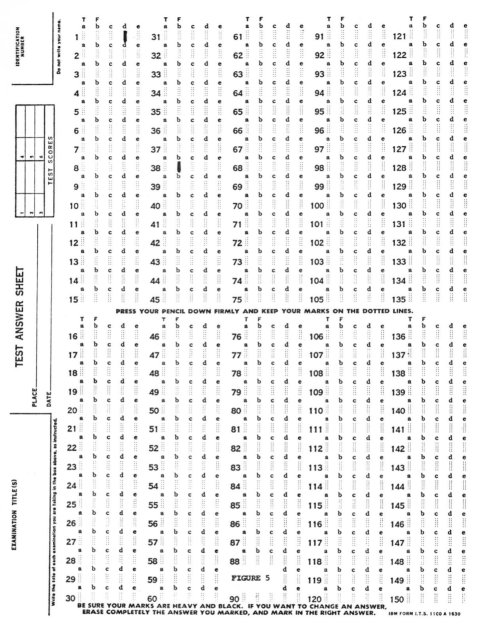

FIGURE 1.4

Assume the first question on the examination is a multiple-choice:

1. Dade is a county in the state of:

   a. California
   b. Virginia
   c. New York
   d. Florida

The answer would be marked alongside 1.d., as shown in Figure 1.4

Assume question number 38 on the examination is a true-false:

38. It is not required to have a certificate of competency to perform building contracting work in Dade County.

The answer would be marked alongside 38.b., as shown in Figure 1.4

If you examine the sample Test Answer Sheet, you will note that the spaces for answers are marked, a, b, c, d, e, or T F. Test Answer Sheets may vary slightly; before you commence your examination take a few moments to study the Sheet and make certain that you follow the sequence of numbering as well as any other instructions that may appear.

SUMMARY

1. Mark your STUDY SCHEDULE up to the maximum time your can afford, and stick to it. Discipline... discipline...discipline...

2. Memorize the 10 RULES FOR EFFICIENT STUDY and make them your way of life until you have secured your certificate.

3. Do not attempt to memorize answers to specific questions. It is not realistic to expect to know everything there is to know about your trade, craft, or profession, and it is no loss of pride to miss some answers. The main objective is to get a passing grade. Part of the strategy of getting a passing grade and receiving your certificate of competency, is understanding this objective and knowing your limitations. Remember, there are many licensed persons who know less than you do.

4. Concentrate your study on those areas in which you know you are weak, and strengthen your "I.Q." around those areas.

5. Knowing where to find the information is half the battle won. Familiarize yourself thoroughly with the table of contents and index of the recommended study material; this is essential for the "open book" section of your examination.

6. Do not study irrelevant material. Your recommended reference list may include some books from which only one chapter--or a short passage--is applicable. For example, if you are taking the examination for well drilling, you will find the book, *Plumbing* by Babbitt, listed. This does not mean that you are responsible for the entire text; confine your studying to those passages that apply to well drilling and piping only. Where volumes of the NFPA are listed, mark out those pamphlets cited, and the important sections referring to your craft, and ignore the remainder.

2. QUESTIONS AND ANSWERS FROM PREVIOUS EXAMS

OPEN BOOK
6 HOUR EXAMINATION

VALUE

Section I	50 points
Section II	25 points
Section III	25 points

MASTER OF REFRIGERATION

SECTION I

Problem Solving
Heat Gain for Walk-In

Given:

1. Application: Meat Market Walk-In Cooler
2. Outside Dimensions: 16 ft wide 16 ft long
   9 ft high
3. Insulation: 4 inch Thickness Cork
4. Glass: Double Glazed Windows 3 ft x 3 ft
5. Overall Wall Thickness: 6 inches
6. Surrounding Air Temperature: 95° F - 50% RH
7. Electrical Load: 600 watts
8. Storage Room Temperature: 35° F
9. Product Load:
   a. 1500 lbs Lean Beef
   b. 1200 lbs Shell Oysters
   c. 2000 lbs Beets

Remarks: The product is brought in at 65 F and is
to be cooled down to 35° F in 15 hours.
Compressor running time is 16 hours.

Solve:

1. The wall load ____a____ Btu/24 hours

   (a) 106347   (c) 122472

   (b) 112862   (d) 139113

2. The air change load _____ Btu/24 hours

   (a) 44,651   (c) 65,463

   (b) 62,921   (d) 71,847

3. The total product load_____ Btu/24 hours

   (a) 169,719   (c) 182,628

   (b) 173,824   (d) 238,680

4. The electrical load_____Btu/24 hours

   (a) 4251   (c) 4080

   (b) 4665   (d) 5467

5. Grand total load_____Bru/24 hours

   (a) 418920   (c) 28133

   (b) 450875   (d) 53211

6. Compressor requirements_____Btu/hr

   (a) 418920   (c) 53211

   (b) 450140   (d) 28179

7. Condensing unit model number required

   (a) C-1   (c) C-3

   (b) C-2   (d) C-4

8. Evaporator unit model number required

   (a) E 5   (c) E 7

   (b) E 6   (d) E 8

Solution:

   Use the ASHRAE Handbook of Fundamentals, 1972,
   pages 446-448, 572-573.

Step 1. Wall load;

```
W 16 ft x L 16 ft x 2 = 512 (Ceiling & Floor)
L 16 ft x H 9 ft  x 2 = 288 (North & South)
L 16 ft x H 9 ft  x 2 = 288 (East & West)
                        ────
                        1088
Minus glass             -  9
Net sq ft               ────
                        1079 ft
```

Step 2. From Table 1, p.446 ASHRAE, select heat gain
         factor "U", 108

         1079 (sq ft) x 108 ("U") = 116532 Btu/24 hrs.

Step 3.   From Table 1, p.446 ASHRAE, select glass
          factor "U", 660

          9 (sq ft) x 660 ("U") = 5940 Btu/24 hrs.

          Total wall & glass load = 122472 Btu/24 hrs.

Step 4.   Air change load;

          Cubical contents = 16 ft L x 16 ft W x 9 ft H
                           = 2304 cu ft

          From Table 4, p.446 ASHRAE, average air
          change is 9.5
          From Table 5, p.447 ASHRAE, heat required
          to cool outside air @ 50% RH is, 2.04
          2304 (cu ft) x 9.5 (air change) x
            2.04 (heat) = 44,651 Btu
          Total air change load = 44,651 Btu/24 hrs.

Step 5.   Product load; temperature reduction is found
          by subtracting storage room temperature from
          the surrounding air temperature 95 F
                                          -35 F
                    Temperature reduction = 60 F
          From Table 4, p.573 ASHRAE; specific heat of
          beef is 0.77, of beets is 0.90, of oysters
          is 0.83
          So, 1500 lbs beef x 0.77 x 60 F   = 69,300
              2000 lbs beets x 0.90 x 60 F   = 108,000
              1200 lbs oysters x 0.83 x 60 F = 59,760

Step 6.   From Table 4, p.572 ASHRAE: heat of
          respiration for beets = 2700 Btu/24 hr/ton
            @ 32 F
          then, $\frac{1200 \text{ (lbs beets) x 2700 (heat)}}{2000 \text{ lbs (1 ton)}}$ = 1620
          Correct for 3 F difference from table;
                1620 x 3 = 4860 Btu/24 hrs.

                            69,300
                           108,000
                            59,760
                             1,620
          Total product load 238,680 Btu/24 hrs.

TABLE 2.1

TYPICAL MANUFACTURER'S RATINGS FOR AIR COOLED
CONDENSING UNIT RATINGS IN BTU/HR

Model	Evaporative Temperature			
	0° F	10° F	20° F	30° F
C 1	19,100	24,800	29,000	38,100
C 2	20,200	25,500	32,300	40,200
C 3	20,600	26,800	34,000	42,300
C 4	30,000	39,200	49,400	61,200

TABLE 2.2

TYPICAL MANUFACTURER'S RATINGS FOR EVAPORATORS

Model	Btu/hr	Cfm	Motor hp	R-12 Charge
E 5	21,200	3,300	2-1/6	15
E 6	26,500	4,200	2-1/6	17
E 7	36,000	6,400	2-1/6	25
E 8	48,300	7,600	2-1/2	50

Step 7. As stated in conditions given, product load must be cooled down in 15 hours.
So, 238680 x $\frac{24 \text{ hrs}}{15 \text{ hrs}}$ = 381,888 Btu
Total product load = 381,888 Btu/24 hrs.

Step 8. Electrical Load (lights on approximately 2 hrs)
600 (watts) x 3.4 (Btu/hr) x 2 (hrs) = 4080

Step 9. Total Btu load per 24 hours =

Wall & glass	122,472
Air change	44,651
Product	238,680
Electrical	4,080
Total Btu load per 24 hours =	409,888

Step 10. Add 10% for safety factor,

	409,888
	40,988
Grand total Btu load per 24 hrs =	450,875

Step 11.
$\frac{\text{Grand total Btu load per 24 hrs}}{\text{Compressor unning time}}$ = Compressor requirements

so, $\frac{450875 \text{ (Btu/24 hrs)}}{16 \text{ (hrs)}}$ = 28179 Btu/hr

Compressor requirements = 28179 Btu/hr

Step 12. From typical Manufacturer's Table select Model C-1 Compressor

Step 13. From typical Manufacturer's Table select Model E-7 Evaporator

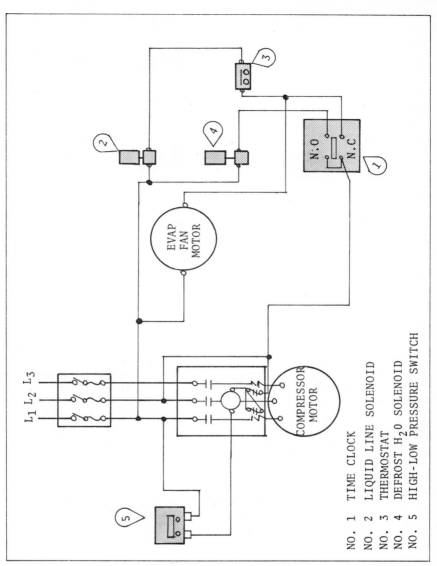

FIGURE 2.1

SECTION II

Wiring diagram (Figure 2.1) for refrigeration control circuit. Answer the following ten (10) questions TRUE or FALSE.

9. The entire control circuit is line voltage.

    (a) T   (b) F

10. When the time clock switch closes on the normally open position it will stop the fan motor while the compressor will continue to run.

    (a) T   (b) F

11. When the thermostat is satisfied it will immediately shut down the compressor.

    (a) T   (b) F

12. The sole function of the time clock is to allow defrosting of the coil.

    (a) T   (b) F

13. When the defrost cycle has been completed by predetermined time setting, the following takes place: The liquid line solenoid open, the H_2O defrost solenoid closes, the evaporator fan motor starts up, and the time clock goes to its normally open position.

    (a) T   (b) F

14. The compressor motor safety is the hi-lo pressure control.

    (a) T   (b) F

15. If conditions make it necessary the compressor will recycle to pump down.

    (a) T   (b) F

16. If the thermostat were removed from the circuit, the evaporator fan motor would run continuously.

   (a) T   (b) F

17. Whenever the evaporator fan motor is not running the H$_2$0 solenoid valve is wide open thus allowing water to circulate across the evaporator coils.

   (a) T   (b) F

18. If the system pressure reaches a point below the setting of the hi-lo pressure switch, the entire system becomes deenergized.

   (a) T   (b) F

SECTION III

19. What control would you use on a cap tube or auto-matic X valve system to maintain desired temper-ature:

   (a) Dual pressure control

   (b) High pressure control

   (c) Temperature control

   (d) Low pressure control

20. What type of electric motor has a set of field coils and a rotating armature:

   (a) Induction motor

   (b) Capacitor split phase

   (c) Repulsion start induction

   (d) Dual winding

21. The total heat of any substance is obtained by adding together its:

   (a) Radiant and specific heat

(b) Latent and sensible heat

(c) Radiant and sensible heat

(d) Specific and latent heat

22. A refrigeration system having two or more circuits each with a condenser and evaporator, where the evaporator of one circuit cools the condenser of the other (lower temperature circuit) is known as:

(a) Multiflow system   (c) Supercool system

(b) Cascade system   (d) Recool system

23. On adding refrigerant to a system through the high side after repairs the compressor suction shut-off valve should be:

(a) Removed   (c) Front seated

(b) Back seated   (d) Closed

24. As the pressure on a liquid rises, the temperature at which the liquid will boil:

(a) Varies   (c) Also rises

(b) Remains the same   (d) Lowers

25. Open motors are generally designed to run at a temperature rise 40 C (72° F). What does this mean?

(a) Indicates the temperature rise over the surrounding air of a motor when running at full name-plate conditions

(b) Indicates the temperature rise of the motor.

(c) Indicates the temperature drop of the surrounding air.

(d) Indicates the temperature drop of the motor at rest

26. A voltmeter is connected into the circuit in which of the following?

    (a) Series   (c) Crossover

    (b) Parallel   (d) Delta

27. All pressure relief devices (not fusible plugs) shall be:

    (a) Directly pressure-actuated

    (b) Remote pressure-actuated

    (c) Not pressure-actuated

    (d) Manually operated

28. The water valve on a refrigeration system should be connected to the motor electrically in _____.

    (a) Series

    (b) Parallel

    (c) Either series or parallel

    (d) It is not connected to motor

29. What causes electric solenoid valves to chatter?

    (a) Sticking valve

    (b) Valve installed reversed

    (c) Low voltage

    (d) Fluctuating pressure

30. The pressure motor control is usually located on the _____.

    (a) Condensing unit   (c) Cabinet

    (b) Motor   (d) Evaporator

31. The low pressure motor control should be connected into the _____.

    (a) Evaporator   (c) Crankcase

(b)   Suction line coil   (c)   Lowest point in
                                      system

32.   The oil separator must be mounted_____.

    (a)   Below compressor

    (b)   Above compressor

    (c)   Under the discharge

    (d)   Level

33.   The separator oil return line must be connected
      to the_____.

    (a)   Liquid receiver

    (b)   Compressor crankcase

    (c)   Low-side oil float

    (d)   Liquid line

34.   In a beverage cooler installation the sweet water
      bath is made of_____.

    (a)   Alcohol   (c)   Tap water

    (b)   Alcohol and brine   (d)   Sugar solution

35.   In a single compressor system the flooded cooling
      coil should be mounted_____.

    (a)   Slight pitch forward

    (b)   Level

    (c)   Pitch to the float

    (d)   Pitch away from float

36.   In a walk-in the average temperature difference
      between the coil refrigerant and the box air is
      _____degrees.

    (a)   10° F   (c)   30° F

    (b)   20° F   (d)   40° F

37. When installing a multiple flooded coil the refrigerant control device is _____.

   (a) TX valve and high-side float

   (b) Auto-X valve and low-side float

   (c) Low-side float and TX valve

   (d) Low-side float high-side float combination

38. The suction pressure reading for refrigerant 12 @ 5° F is:

   (a) 19.5   (c) 11.8

   (b) 6.5   (d) 2.8

MASTER OF REFRIGERATION

ANSWER SHEET

Section I

1 - C
2 - A
3 - D
4 - C
5 - B
6 - D
7 - A
8 - C

Section II

9 - T
10 - T
11 - F
12 - T
13 - F
14 - F
15 - T
16 - F
17 - T
18 - F

Section III

19 - C
20 - A
21 - B
22 - B
23 - B
24 - C
25 - A
26 - B
27 - A
28 - B
29 - D
30 - A
31 - C
32 - D
33 - A
34 - C
35 - B
36 - B
37 - C
38 - C

MASTER OF STEAM GENERATING BOILERS AND HEATING

CLOSED BOOK
4 HOUR EXAMINATION

MASTER OF STEAM GENERATING BOILERS AND HEATING

1. Which of the following is not a type of steam trap?

   (a) Bucket trap

   (b) Float and thermostatic trap

   (c) P-trap

   (d) Impulse trap

2. Which of the following is a type of steam heating system?

   (a) Direct expansion   (c) Reverse return

   (b) Vacuum   (d) Injection

3. In the discussion of oil heating using a fuel pump, the number of pounds per square inch pump pressure must drop to close nozzle valve is known as:

   (a) Delivery   (c) Head of oil

   (b) Valve differential   (d) Lift

4. If a fuel system is being checked in which the tank is buried below the burner level and the oil passes through a line filter, the _____ must show a reading; if it does not, an air leak is present.

   (a) Pressure gauge   (c) Vacuum gauge

   (b) Fuel gauge   (d) Sight glass

5. What type of boiler has water in the tubes being heated by the hot gases outside the tubes?

   (a) Water tube boiler   (c) Radiant boiler

   (b) Fire tube boiler   (d) Atmospheric boiler

6. Which one of these methods of heat transfer is not employed by the various types of heat transfer units?

(a) Radiation

(c) Gravitation

(b) Convection

(d) Conduction

7. Which of the following is considered a heating accessory?

(a) Boiler

(c) Coils

(b) Piping

(d) Pumps

8. Which of the following is not a MAIN part of a heating system?

(a) Boiler

(c) Heat transfer surfaces

(b) Piping

(d) Water

9. Which one of the following is a reason for a steam boiler getting too much water?

(a) Leaking hot water coil in boiler

(b) Faulty swing check in return header

(c) Trap plugged up

(d) Return pump not operating

10. What is the weight of one cubic foot of water at 50 feet?

(a) 40.52 lbs

(c) 60.33 lbs

(b) 62.41 lbs

(d) 70.01 lbs

11. When the pitch of the thread on a pipe is 1/4", how many turns are required to thread 2-1/2" of the pipe?

(a) 8

(c) 12

(b) 10

(d) 13

12. If a fuel oil unit has been installed with too small tubing on high lifts and long runs, what will the result most probably be?

    (a) Cause a squirt out of the nozzle

    (b) It will read a vacuum on the lift instead of pressure

    (c) The oil will have a milky appearance due to air being drawn with it

    (d) It will cause the oil to separate

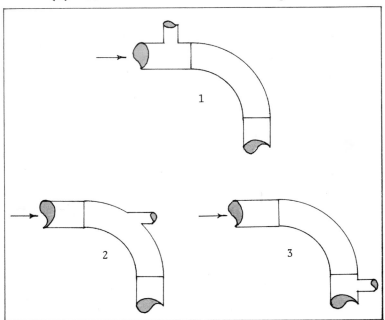

13. Which of the above Ports will measure TRUE STATIC PRESSURE?

    (a) Port No. 1   (b) Port No. 2   (c) Port No.3

14. What is the purpose of an expansion joint?

    (a) To minimize vibration

    (b) To allow for normal movement of the pipe due to expansion and contraction

    (c) To compensate for movements of pumps or compressors

    (d) Make removal of pipe easy

15. What procedure should be followed when a beam interferes with the installation of a pipe line and the piping cannot be dropped below the beam and has to be looped over the beam:

    (a) Provide for an anchor to prevent vibration

    (b) Provide for venting air from high point of pipe

    (c) Use oversized fittings and pipe to prevent restriction

    (d) Make sure that expansion joint is used

16. The treatment of liquids under pressure and at rest is:

    (a) Hydrostatics   (c) Hydrophonics

    (b) Hydraulics   (d) Compression

17. The treatment of liquids under pressure and in motion is:

    (a) Hydraulics   (c) Hydrophonics

    (b) Hydrostatics   (d) Compression

18. A pipe run is 81.375 feet long. If divided into 21 equal parts, how long will each division be?

    (a) 3.875 feet   (c) 4.015 feet

    (b) 3.750 feet   (d) 4.090 feet

19. What type valve would be used in a horizontal line that requires complete drainage?

    (a) Globe   (c) Gate

    (b) Check   (d) Relief

20. Which of the following valves should not be used for throttling purposes?

    (a) Plug   (c) Gate

    (b) Needle   (d) Globe

21. On flanged joints, bolts should be tightened in what manner:

    (a) Cross over method (c) By hand only

    (b) Rotation   (d) Welded

22. All piping systems shall be capable of withstanding a hydrostatic test pressure of how many times the designed pressure:

    (a) 3   (c) 2

    (b) 1-1/2   (d) 2-1/2

23. A circumferential weld in pipe, fusing the abutting pipe walls completely from inside wall to outside wall is called:

    (a) Collet weld   (c) Jay weld

    (b) Butt weld   (d) Fillet weld

24. Which of the following is not generally used to take care of thermal expansion in pipe lines?

    (a) Loop in the line

    (b) Packed slip-joints

    (c) Bellows type joints

    (d) Change in size of line

25. Anchors are not required on pipe lines subject to expansion and contraction in order to:

    (a) Reduce thermal expansion

    (b) Cause movement due to the thermal expansion, to take place at desired points

    (c) Prevent damage to pressure vessels

    (d) Prevent damage to building structure

26. The term "induced draft" could refer to:

    (a) A type of control   (c) A type of diagram

    (b) A type of boiler   (d) A type of compressor

27. Air in oil line to burner may cause which one of the following:

    (a) Pulsation

    (b) Excessive fuel consumption

    (c) Blue flame

    (d) Dirty nozzle

28. A pipe which conducts condensation from the supply side to the return side of a steam heating system is:

    (a) Drip   (c) Runout

    (b) Riser   (d) Hartford connection

29. A pop safety valve on a boiler is usually provided with a hand lever with its purpose being:

    (a) For closing the valve once it pops open

    (b) To relieve air

    (c) For manual safety

    (d) For testing the valve

30. Steam mains and returns should be pitched not less than one inch per_____feet in the direction of the steam flow.

    (a) 10 feet   (c) 30 feet

    (b) 20 feet   (d) 40 feet

31. Each steam boiler shall have at least how many water glasses:

    (a) 0   (c) 2

    (b) 1   (d) 3

32. A typical control installation for a gas-fired steam or gravity system, with a two-wire solenoid, all the control units are connected _____ so that opening of the contacts in the thermostat, the limit control on the pilot-safety control breaks the power circuit to the gas valve and allows the valve to close.

    (a) In series

    (b) In parallel

    (c) In series or parallel

    (d) Through a relay

33. The capacity for storing heat, varying with the mass of material and the specific heat is called:

    (a) Charge capacity   (c) Thermal capacity

    (b) Storage capacity   (d) Mass capacity

34. Valve which goes to the closed position when power, electric or pneumatic, is cut off:

    (a) Normally positioned valve

    (b) Normally open or closed valve

    (c) Normally open valve

    (d) Normally closed valve

35. When a hydrostatic test is applied to fittings and appliances on a boiler, the required test pressure by code shall be_____the maximum allowable working pressure.

    (a) 1-1/2 times   (c) The same as

    (b) 2 times   (d) Not lower than

36. Each boiler shall be provided with a valve connection at least_____pipe size for the exclusive purpose of attaching a test gauge when the boiler is in service, so that the accuracy of the boiler steam gauge can be ascertained.

    (a) 1/8"   (c) 1/2"

    (b) 1/4"   (d) 3/4"

37. A packaged steam boiler's rated capacity in lbs. of steam per hour (212°F) is 1380. Boiler hp is:

    (a) 30   (c) 45

    (b) 40   (d) 35

38. The part of a return main of a steam heating system which is filled with water of condensation is a:

    (a) Return header   (c) Main return

    (b) Dry return   (d) Wet return

39. A casting or an assembly of pipe fittings which provides a seal between a horizontal return main and a vertical connection to another return main at a higher level is a:

    (a) Spring-piece or runout

    (b) Riser or vertical pipe

    (c) Lift connection or lift fitting

    (d) Return header or horizontal piping

40. On power boilers, fusible plugs which are inserted from the water side of the plate, flue or tube to which they are attached are known as:

(a) Outside plugs   (c) Fire-side plugs

(b) Water-side plugs   (d) Inside plugs

41. Steam heating boilers: The maximum allowable
     working pressure shall not exceed_____psi
     on a boiler used exclusively for low pressure
     steam heating.

     (a) 10   (c) 5

     (b) 20   (d) 15

42. Each steam-discharge outlet, except safety valve,
     reheater inlet and outlet, or superheater inlet
     connections, shall be fitted with a _____.

     (a) Gauge   (c) Stop valve

     (b) Check valve   (d) Balancing cock

43. A boiler operating at a pressure greater than
     100 psi, the blow-off piping thickness shall not
     be less than:

     (a) Schedule 80 pipe

     (b) Schedule 40 pipe

     (c) Standard black iron pipe

     (d) Heavy copper pipe

44. For boilers having a heating surface of not more
     than 100 sq ft, the feed water pipe shall not be
     less than:

     (a) One-half inch pipe size

     (b) Two 3/8" pipes

     (c) One inch pipe size

45. The feed pipe shall be provided with:

     (a) Two check valves, one of which is near
          the boiler

     (b) A check valve near the boiler and a valve
          or cock between the boiler and the check
          valve

    (c)   Two check valves with globe valves between them

46. Each steam boiler shall have at least one water glass, the lowest visible part of which shall be not less than:

    (a)   Four inches above the lowest permissible water level

    (b)   The tube level

    (c)   Two inches above the lowest permissible water level

47. A standard of installation of boilers of three horsepower or less, generating steam at less than 100 psi pressure, shall be to provide clearance or insulation in such a manner that continued operation of the boiler will not raise the temperature of surrounding combustible construction above:

    (a)   170° F   (c)   40° Centigrade

    (b)   Normal temperature

48. The steam pressure gauge must be connected to:

    (a)   The steam space of the boiler

    (b)   The main steam line

    (c)   The water space of the boiler

49. What is required between the steam gauge and the boiler:

    (a)   A trap or syphon with a valve off the tee or lever handle type

    (b)   A vacuum breaker check valve

    (c)   Cross fittings to facilitate cleaning

50. When a globe valve is used in the feed water line:

    (a)   The valve must be installed so that the boiler pressure is beneath the valve disc so the valve closes against boiler pressure

(b) The valve must be installed so that the feed water pressure is beneath the valve disc so the valve closes against feed pressure

(c) Either of the two ways

51. A conventional float-operated low water fuel cut-off will turn off_____but will not mechanically feed water to a boiler.

(a) A thermostatic fuel burner

(b) An automatic fuel burner

(c) An automatic water feeder

(d) An electric water feeder

52. A steam trap is an automatic valve used in steam systems and its total function is which of following:

(a) Permits passage of condensate

(b) Permits passage of air

(c) Permits passage of condensate, air and non-condensable gases

(d) Permits passage of steam and air

53. A 100 hp boiler evaporates 3450 lbs. of steam per hour. Find size of the condensate receiver knowing the following facts:

75% of gross volume of tank is usable. Receiver to hold volume equal to condensate evaporated by boiler in a one half hour period at normal firing rate of boiler. The size of the receiver is:

(a) 352 gal.   (c) 154 gal.

(b) 276 gal.   (d) 510 gal.

54. A low pressure 25 hp steam boiler shall be enclosed with which of the following type walls:

    (a)   Any type of wall

    (b)   2 hour fire-resistive construction

    (c)   3 hour fire-resistive construction

    (d)   4" Concrete block walls

55. A packaged hot water boiler has a rated capacity of 640,000 Btu per hour. Boiler is fired with natural gas - Boiler is 80% efficient - Fuel consumption in therms per hour is:

    (a)   6   (c)   8

    (b)   12   (d)   10

56. In a hot water heating system using a closed expansion tank, vents are required at:

    (a)   Expansion tank only

    (b)   Discharge side of pump

    (c)   Last coil in system

    (d)   All high points

57. A type "B" flue is proposed for a natural gas fired steam boiler. It may *not* be used, according to South Florida Building Code if flue gas temperature is in *excess* of:

    (a)   350° F   (c)   500° F

    (b)   400° F   (d)   450° F

58. In an automatically fired steam boiler the lowest safe water line with reference to water glass should be:

    (a)   Top of water glass

    (b)   Midway in water glass

    (c)   No lower than lowest visible part of the water glass

    (d)   3/4 up from bottom of water glass

59. What is the type of boiler with water in tubes and gases of combustion flowing around outside of tubes?

    (a)   Fire tube boiler

    (b)   Radiant boiler

    (c)   Indirect fire tube boiler

    (d)   Water tube boiler

60. Comfort heating system in a building requires a 810,00 Btu per hour heating coil - Circulating hot water is to be used - Incoming water to coil is 180° - Water leaving coil is 152° - Gpm of circulated water to handle load is:

    (a)   90 gpm   (c)   86 gpm

    (b)   100 gpm   (d)   122 gpm

61. A slow pick up of the lead from a "cold" boiler start will:

    (a)   Improve combustion efficiency

    (b)   Decrease stack temperature 200° or more

    (c)   Lengthen life of boiler

    (d)   Shorten life of boiler

62. A warm air furnace is sized for a space requiring 112,00 Btu/hr - The temperature rise of supply air over return air is 100° cfm (nearest) required for this conditions is:

    (a)   1020 cfm   (c)   1100 cfm

    (b)   870 cfm   (d)   1210 cfm

63. In an automatically fired steam boiler the lowest safe water line with reference to water glass should be:

    (a)   1/3 up from bottom of water glass

    (b)   No lower than lowest visible part of the water glass

(c) Top of water glass

(d) Midway in water glass

64. The minimum vent size allowed by code for oil storage is:

(a) 1"

(c) 1-1/4"

(b) 1-1/2"

(d) 2"

65. All branch (take off) lines from a steam main should be taken off from:

(a) Side of main

(c) Bottom of main

(b) Top of main

(d) Top or bottom of main

66. An A.S.M.E. Pressure relief valve on a boiler is usually provided with a hand lever. Its primary purpose is to:

(a) Close valve once it has opened

(b) To change operating pressure

(c) To test valve and relief of air on initial fill

(d) Provide manual safety

67. According to Code Certificate of Inspection for High Pressure Steam Boiler shall be for a fixed period not exceeding:

(a) 1 year

(c) 6 months

(b) 3 months

(d) 9 months

68. On starting up a boiler using No. 2 fuel oil a flue gas analysis shows 7% CO_2 - How would this CO_2 reading be rated?

(a) Excellent

(c) Fair

(b) Good

(d) Poor

69. One important function of primary control on oil fired furnace is to:

 (a) Open oil solenoid valve

 (b) Stop operation of unit in case of fire

 (c) Stop burner operation if flame is not established or if flame is extinguished after being in operation

 (d) Control oil pressure at burner nozzle

70. Open expansion tanks are open to atmosphere and are located:

 (a) Anywhere in piping system

 (b) Discharge side of pump

 (c) Suction side of pump above highest unit in system

 (d) At heating cooling change over valve

71. For adjusting pressure drop thru a heating coil the valve to be used is:

 (a) Three way mixing valve

 (b) Plug cock valve

 (c) Gate valve

 (d) Needle valve

72. Air in supply oil line to burner may cause which of following:

 (a) Blue flame

 (b) Excessive fuel consumption

 (c) Dirty nozzle

 (d) Pulsation and singing noise

73. A noisy heating system may be caused by _____.

   (a) Piping pockets

   (b) Boiling water in an open system

   (c) Floor holes too small for risers

   (d) A B & C

74. On an undersized job which is delivering insufficient heat, _____would *not* help to correct the condition.

   (a) Painting radiators bronze

   (b) Installing copper coils in fire pot

   (c) Increase H_2O temperature limit control

   (d) Increase the air at the vacuum pump air ejector

75. Correct safety control requires an ASME relief valve on _____.

   (a) All high pressure steam systems

   (b) On all low pressure steam systems

   (c) On all ASME rated systems

   (d) On every hot water heating boiler

76. An ASME relief valve should be connected _____.

   (a) To the top of the boiler

   (b) To the supply main at the boiler

   (c) Ahead of the fill-valve

   (d) On the return main downstream of the pump

77. The function of a *low water cut-off* is:

   (a) To shut off the water feeder

   (b) To open the fill valve and allow the water to rise

(c)   To cut off the pump on the return main

(d)   To open the electrical circuit on the
auto-burner

78.   On the average hot water boiler the maximum working
pressure for which the relief valve is set, is____psi.

(a)   12   (c)   65

(b)   30   (d)   120

79.   To correct heating application practice requires a
low water cut-off on _____.

(a)   All hot water boilers

(b)   All high pressure steam boilers

(c)   All boilers with automatic water feed

(d)   All steam and hot water boilers

*The Following 5 Questions Pertain to Steam and Vapor
Heating Systems.*

80.   Units heat too slowly or only partially through a
building may be caused by _____.

(a)   Foaming boiler

(b)   The isn't any Hartford loop on the return

(c)   Defective traps

(d)   A B & C

81.   System heats building well but one or several rooms
get insufficient heat. Condition may be caused by
_____.

(a)   Inefficient pump

(b)   Steam traps blow through

(c)   Steam trap stuck shut

(d)   A B & C

82. If a system is not functioning properly due to a flooded radiator main the proper remedy would be to _____.

    (a) Pitch radiator in proper direction

    (b) Throttle the by-pass valve

    (c) Install a Hartford equalizer loop on return

    (d) Adjust pressure control to higher setting.

83. On a two-pipe system malfunction caused by trapped radiators, it is most likely that _____.

    (a) The radiator return is higher than the inlet

    (b) Boiler water line is higher than end of main

    (c) The radiators are improperly vented

    (d) High boiler water has decreased the steam volume

84. A system will not hold vacuum, which of the following would *not* cause the condition?

    (a) Steam traps blow through

    (b) Bad pump

    (c) Defective air eliminators

    (d) Steam trap stuck shut

85. The symbol Ⓧ in a piping system indicates _____.

    (a) Expansion valve

    (b) Strainer

    (c) Thermostatic trap

    (d) Check valve

THE FOLLOWING 15 QUESTIONS ARE TO BE ANSWERED TRUE OR FALSE.
IF TRUE MARK T IF FALSE MARK F:

86. Low voltage controls are usually designed to operate
     at not over 25 volts.

                    T.   F.

87. On three phase controls a magnetic switch with a
     built-up holding coil must be used.

                    T.   F.

88. A change in space temperature will cause the limit
     control to automatically fire a heating system.

                    T.   F.

89. When a line voltage thermostat is used to control a
     hot water circulating pump, a relay is unnecessary
     unless the motor load exceeds the stat capacity.

                    T.   F.

90. When a line voltage hot water control is sensing
     storage tank temperature, and controlling a single
     phase, fractional horsepower, booster pump to
     circulate boiler water through a low heater; it is
     not required to fuse the circuit.

                    T.   F.

91. On a 2-floor, 2-zone hydronic system, two space
     stats and two motorized valves are required.

                    T.   F.

92. A primary control may be;

          1. Stack mounted
          2. Burned mounted with stack sensor
          3. Burner mounted with air tube sensor
          4. Burner mounted with light-sensitive flame
              detector

                    T.   F.

93. The final control to act in a gas heating system is the pilot safety.

    T.   F.

94. In oil fired heat systems, the first control to act is the primary.

    T.   F.

95. Electric heating systems are thermally operated sequence controls or thermally operated staging controls as time-delay relay to energize the heating elements.

    T.   F.

96. Split-phase motors are only available in fractional horsepower; they use a centrifugal system which disconnects the starting windings as soon as the motor reaches operating speed.

    T.   F.

97. A *capacitor-start motor* develops a lower starting torque than a *permanent-split capacitor motor*.

    T.   F.

98. A gauge-glass low-water cutoff is a device to prevent the water in a steam boiler from falling below a safe level. It is essentially an electric control operated by a mercury switch, and de-energizes the burner.

    T.   F.

99. To maintain minimum discharge temperature, the space thermostat should override the low limit discharge in the air stream.

    T.   F.

100. Electrically operated circuit breakers sometimes are used for starters with very large squirrel cage induction motors.

    T.   F.

MASTER OF STEAM GENERATING BOILERS AND HEATING

ANSWER SHEET

1 - C	26 - B	51 - B	76 - A
2 - B	27 - A	52 - C	77 - D
3 - B	28 - A	53 - B	78 - B
4 - C	29 - D	54 - C	79 - D
5 - A	30 - B	55 - C	80 - D
6 - C	31 - B	56 - D	81 - C
7 - D	32 - A	57 - C	82 - C
8 - D	33 - C	58 - C	83 - A
9 - A	34 - D	59 - D	84 - D
10 - B	35 - A	60 - A	85 - C
11 - B	36 - B	61 - C	86 - T
12 - D	37 - B	62 - A	87 - T
13 - C	38 - D	63 - B	88 - F
14 - B	39 - C	64 - C	89 - T
15 - B	40 - B	65 - B	90 - F
16 - A	41 - D	66 - C	91 - T
17 - A	42 - C	67 - C	92 - T
18 - A	43 - A	68 - D	93 - F
19 - C	44 - A	69 - C	94 - F
20 - C	45 - B	70 - C	95 - T
21 - A	46 - C	71 - B	96 - T
22 - B	47 - A	72 - D	97 - F
23 - D	48 - A	73 - D	98 - T
24 - D	49 - A	74 - D	99 - F
25 - A	50 - B	75 - D	100 - T

MASTER OF SERVICE AND MAINTENANCE
(Refrigeration, air-conditioning, and heating)

CLOSED BOOK
4 HOUR EXAMINATION

1. When using a double element time delay fuse with a 20 amp. motor, the maximum fuse size should be:

    (a) 60 amps.   (c) 30 amps.

    (b) 20 amps.   (d) 25 amps.

2. A two-stage refrigeration system should be used when:

    (a) Ammonia is used as the refrigerant

    (b) Compression ratio would be very high

    (c) Compression ratio would be very low

    (d) Suction temperature would be very high

3. A double riser suction line is used to:

    (a) Allow refrigerant vapor to return more easily

    (b) Cause lower velocity of refrigerant vapor

    (c) Allow the compressor to pump more cfm

    (d) Allow more satisfactory oil return

4. On a motor a running capacitor is used to:

    (a) Increase torque

    (b) Increase horsepower

    (c) Increase power factor

    (d) Increase amperage

5. Fresh meat will keep better when stored in an area with:

    (a) 50% relative humidity

    (b) 65% relative humidity

    (c) 85% relative humidity

    (d) 99% relative humidity

6. The bleed-off from a water tower (cooling tower) should be:

    (a) Equal to the amount evaporated from the tower

    (b) Half the amount evaporated from the tower

    (c) Double the amount evaporated from the tower

    (d) 1 gpm per ton

7. With a given size condensing unit and given refrigerator temperature more moisture will be removed from the air of the refrigerator if the evaporator is operated:

    (a) 5° F below refrigerator temperature

    (b) 15° F below refrigerator temperature

    (c) 20° F below refrigerator temperature

    (d) 5° F above dew point temperature of the air in the refrigerator

8. A mechanic measures a velocity of 600 ft per minute discharging from a 30 x 6 wall grille with a 90% free area. Which of the following is the nearest to actual air quantity emerging from grille:

    (a) 600 cfm   (c) 800 cfm

    (b) 650 cfm   (d) 815 cfm

9. A modulation thermostat employs what to achieve its purpose:

    (a) Pressure

    (b) Transformer

    (c) Switch contacts

    (d) A potentiometer winding

10. In the discussion of the refrigeration cycle, Compression Ratio is expressed in the following equation thus:

(a) Compression ratio - $\dfrac{\text{Discharge pressure psi absolute}}{\text{Suction pressure psi absolute}}$

(b) Compression ratio - $\dfrac{\text{Suction pressure psi absolute}}{\text{Discharge pressure psi absolute}}$

(c) Compression ratio - $\dfrac{\text{Discharge temperature absolute}}{\text{Suction temperature absolute}}$

(d) Compression ratio - $\dfrac{\text{Suction temperature absolute}}{\text{Discharge temperature absolute}}$

11. Pressure limiting devices shall be provided on all systems operating above atomospheric pressure and containing more than how many pounds of refrigerant:

(a) 100 lbs   (c) 20 lbs

(b) 75 lbs   (d) 50 lbs

12. A system having two or more frigerant circuits each with a condenser and evaporator where evaporator of one circuit cools the condenser of the other (lower temp.) circuit, is called a:

(a) Intercooling system

(b) Counterflow system

(c) Cascade system

(d) Two-step system

13. What are the suggested possibilities of trouble indicated by the gauge and superheat readings; low suction pressure - low superheat:

(a) Overcharge of refrigerant

(b) Improper superheat adjustment

(c) Compressor undersized

(d) Evaporator oil-logged

14. On refrigerant piping single compressor systems, what maximum pressure drop should be used in sizing liquid lines:

    (a) 10 psi   (c) 1 psi

    (b) 6 psi   (d) 3 psi

15. A grille equipped with a damper or control valve is a:

    (a) Vane   (c) Damper

    (b) Register   (d) Louvre

16. On refrigeration systems all pressure relief devices (not fusible plugs) shall be:

    (a) Directly pressure-actuated

    (b) Remote pressure-actuated

    (c) Not pressure-actuated

    (d) Manually operated

17. A unit of power equal to one horsepower delivered at the shaft of an engine or motor is commonly known as:

    (a) Amperage   (c) Voltage

    (b) Brake horsepower   (d) Horsepower

18. A system in which both the gas pressure and the oil level in all refrigeration compressors, connected in multiple, are equalized is known as:

    (a) Equalizer crankcase double pipe system

    (b) Equalizer crankcase single pipe system

    (c) Equalizer external system

    (d) Equalizer internal system

19. Horsepower is what unit of power:

    (a) The effort necessary to raise 36,000 lbs
         a distance of one foot in one minute

    (b) The effort necessary to raise 33,000 lbs
         a distance of one foot in one minute

    (c) The effort necessary to raise 24,000 lbs
         a distance of one foot in one minute

    (d) The effort necessary to raise 12,000 lbs
         a distance of one foot in one minute

20. The pressure due to the weight of a liquid in a
    vertical column or more generally the resistance due
    to lift is called:

    (a) Static head   (c) Hydraulic pressure

    (b) Head pressure   (d) Temperature head

21. What is the purpose of a thermostat:

    (a) To control the heat

    (b) To control the cold

    (c) To cycle the cooling system

    (d) To control the temperature

22. What is the purpose of the holding circuit in a
    magnetic motor starter:

    (a) Holds the main contacts open until the control
         circuit through the interlock is made

    (b) Holds the main contacts closed until the
         control circuit through the interlock is broken

    (c) Holds the main contacts open until the control
         circuit through the interlock is broken

    (d) Holds the main contacts closed until the
         control circuit through the interlock is made

23. Open motors are generally designed to run at a temperature rise of 40° C or 72° F, what does this mean:

    (a) Indicates the temperature rise over the surrounding air of a motor when running at full nameplate conditions

    (b) Indicates the temperature rise of the surrounding air

    (c) Indicates the temperature drop of the surrounding air

    (d) Indicates the temperature rise over the surrounding air of a motor when not running

24. In a direct expansion chiller the refrigerant is on what side of the tubes:

    (a) On the shell   (c) Outside

    (b) Inside   (d) Both sides

25. The total minimum area of the openings in the air outlet or inlet through which air can pass is known as what area:

    (a) Free area   (c) Total area

    (b) Core area   (d) Drop area

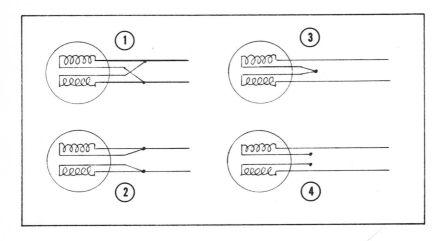

26. From the above diagram for 110-220 volt motor windings, which is the correct hook up for 110 volts?

    (a) 1   (c) 3

    (b) 2   (d) 4

27. From the above diagram for 110-220 volt motor windings, which is the correct hook up for 220 volts?

    (a) 1   (c) 3

    (b) 2   (d) 4

28. If a thermostatic expansion valve is only partially feeding a coil, it is said to be:

    (a) Operating at a low superheat

    (b) Operating with no superheat

    (c) Superheated

    (d) Operating at a high superheat

29. If three unmarked gas cylinders R-12, R-22 and R-500 respectively are all about one-half full and have been stored in the same room for several days at a temperature of 80° F which cylinder contains the R-22:

    (a) There is no way to tell if not marked

    (b) The cylinder that gives a gauge reading of 103#

    (c) The cylinder that gives a gauge reading of 145#

    (d) The cylinder that gives a gauge reading of 84#

30. Heat being a form of energy cannot be:

    (a) Measured   (c) Destroyed

    (b) Felt   (d) Transferred

31. Mercury is used in thermometers because it:

    (a) Expands uniformly

    (b) Does not freeze until it reaches 38.2° F

    (c) Does not boil until it reaches 674.6° F

    (d) Does all of the above

32. Removal of heat from compressed gas between compression stages is known as:

    (a) Recooling   (c) Subcooling

    (b) Intercooling   (d) Upcooling

33. What will the readings of wet and dry hygrometers be if exposed to air that is completely saturated:

    (a) They will both read alike

    (b) The wet bulb will be lower

    (c) The dry bulb will be lower

    (d) 90% relative humidity

34. Type K copper tube has a greater wall thickness than Type L copper tube:

    (a) True   (b) False

35. What type of electric motor has a set of field coils and a rotating armature:

    (a) An induction motor

    (b) A capacitor split phase motor

    (c) A repulsion start indiction run motor

    (d) Dual winding repulsion induction motor

36. When air is passed through a water spray some of the moisture of the air is given up only if the temperature of the water is:

    (a) Below the dewpoint of the air

    (b) Below the temperature of the air

    (c) Above the dewpoint of the air

    (d) Above the temperature of the air

37. On a magnetic starter the alloy piece holding an overload relay closed which melts when current drawn is too great, is known as the thermal overload--

    (a) Holding coil   (c) Relay

    (b) Heater   (d) Element

38. What factors determine the size of an expansion tank:

    (a) The amount of space the water in the system requires in its expanded state

    (b) The amount of space the air in the system requires in its expanded state

    (c) The amount of air in system

    (d) The operating pressure in system

39. What is the gross weight of refrigerant cylinder:

    (a) The weight of the refrigerant plus the weight of the cylinder

    (b) The weight of the cylinder

    (c) The weight of the contents only

    (d) The weight of cylinder and line

40. What does an anemometer measure?

    (a) Feet   (c) cu ft

    (b) Feet per minute   (d) cu ft per minute

41. The term "induced draft" could refer to:

    (a) A type of control

    (b) A type of compressor

    (c) A type of cooling tower

    (d) A type of diagram

42. What fitting should be used to connect 1/4" O.D. copper tubing to a 1/4" internal pipe thread opening in the compressor body?

    (a) A union   (c) A tee fitting

    (b) A half-union   (d) A street ell

43. A non-positive displacement compressor:

   (a) Positive   (c) Centrifugal

   (b) Rotary   (d) Reciprocating

44. Device for removing dust from the air by means of electric charges induced on the dust particles:

   (a) Electric precipitator

   (b) Electric washer

   (c) Electric magnet

   (d) Electric ejector

45. A refrigerant evaporator in which the heat transfer surface is immersed in the refrigerant being evaporated:

   (a) Condenser

   (b) Dry type evaporator

   (c) Flooded evaporator

   (d) TX coil evaporator

46. The gas resulting from the instantaneous evaporation of refrigerant in a pressure-reducing device to cool the refrigerant to the evaporation temperature obtained at the reduced pressure:

   (a) Superheated gas   (c) Inert gas

   (b) Flash gas   (d) Non-condensible gas

47. Latent heat involved in the change between liquid and vapor states:

   (a) Heat of fusion

   (b) Heat of vaporization

   (c) Heat of reaction

   (d) Heat of the liquid

48. The ratio of the weight of water vapor associated with a pound of dry air to the weight of water vapor associated with a pound of dry air saturated at the same temperature:

    (a) Specific humidity   (c) Absolute humidity

    (b) Relative humidity   (d) Percentage humidity

49. An instrument for measuring pressure, essentially a U-tube partially filled with a liquid, usually water mercury, or a light oil, so constructed that the amount of displacement of the liquid indicates the pressure being exerted on the instrument:

    (a) Potentiometer   (c) Manometer

    (b) Volometer   (d) Anemometer

50. A piping system in which the heating or cooling medium from several heat transfer units is returned along paths arranged so that all circuits composing the system or composing a major subdivision of it are of practically equal length.

    (a) Reversed return   (c) Two pipe

    (b) Indirect   (d) Down feed

51. On a steam system what valve should be used in a horizontal line where drainage is required?

    (a) Any valve   (c) Gate valve

    (b) Globe valve   (d) Check valve

52. How much should the overall drop (pitch) be for a 50 foot steam main?

    (a) 2-1/2"   (c) 1"

    (b) 3"   (d) 5"

53. The lowest safe water line in a boiler should not be lower than what part of the gauge glass:

    (a) When no water is visible

    (b) Lowest visible part

    (c) Highest visible part

    (d) Middle part

54. On water pipes an eccentric reducer should be used on which of the following:

    (a) Horizontal Pipes   (c) Vertical pipes

    (b) Brass pipes   (d) Wrought iron pipes

55. When condensate backs up in the return lines because of lack of proper head between the dry return and boiler water level, the water line in a steam boiler will:

    (a) Rise several inches (c) Not vary

    (b) Rise   (d) Drop

56. The purpose of a Hartford Loop is to:

    (a) Prevent backing water out of a boiler

    (b) Remove air from the return line

    (c) Allow for pipe expansion

    (d) Provide a balance for the steam header

57. On a steam system the operation of an inverted bucket trap is based on the:

    (a) Rise and fall of a float

    (b) Combination of steam pressure and weight of the condensate

    (c) Steam pressure drop

    (d) Weight of the water in the trap

58. Which one of the following is not a common type of steam heating system:

   (a) Vapor

   (b) Atmospheric

   (c) Vacuum

   (d) One pipe gravity

59. What type boiler has water around the outside of the tubes being heated by the hot gases within the tubes:

   (a) Water tube boiler

   (b) Fire tube boiler

   (c) Radiant boiler

   (d) Atmospheric boiler

60. Where oil fuel tanks are lower than the burner on fuel burning equipment, it is recommended that they have what kind of piping system:

   (a) Plastic pipe system

   (b) 3 pipe system

   (c) 2 pipe system

   (d) 1 pipe system

61. Which one of the following is a type of steam heating system:

   (a) Direct expansion

   (b) Vacuum

   (c) Reverse return

   (d) Injection

62. Whenever air flows through a pipe or duct, some pressure is lost because of friction, hence the power required to deliver a given quantity of air _____ as the size of the duct is decreased.

   (a) Remains the same

   (b) Increases

   (c) Decreases

63. If air or other noncondensible gases are present in the condenser, the head pressure may rise to a point considerably above the pressure corresponding to the temperature at which the vapor is condensing. One way of determining whether or not there is air in the system is to allow the compressor to stand idle long enough for it, and all other parts of the refrigerating system to cool down to the temperature of the surrounding air. The water supply to the condenser should be shut off, and it is desirable to drain all the water from the condenser. After the entire system has cooled to the temperature of the surrounding air, the reading

of the high pressure should not be more than about
_____ above the pressure corresponding to the
surrounding air temperature.

    (a) 5 psi   (c) 0 psi

    (b) 20° F   (d) 10 psi

64. How many pounds of water can be heated 6 degrees by 72 Btu?

    (a) .833   (c) 432

    (b) 12   (d) 72

65. The total heat content of humid air depends upon its_____.

    (a) Wet bulb temperature alone

    (b) Dry and wet bulb temperature

    (c) Dry bulb temperature

    (d) Percentage of humidity

66. The actual outside diameter of all copper piping is 1/8" larger than the nominal pipe size:

    (a) True   (b) False

67. On a R-12 system if a TX valve were adjusted for a superheat of 10 degrees, it would admit only enough liquid to the coil to maintain the boiling temperature of the liquid inside the coil at _____ as long as the vapor leaving the coil was at a pressure of 46.8 psig.

    (a) 40° F   (c) 45° F

    (b) 50° F   (d) 55° F

68. When TX thermal bulbs are securely attached to the outside of the suction pipe they need not be insulated.

    (a) True   (b) False

69. On a system in operation, a warm liquid line without high head pressure is an indication of:

    (a) Shortage of condenser water

    (b) Expansion valve closed

    (c) Air in condenser

    (d) Shortage of refrigerant

70. In practically all coils used for cooling air, the flow of air and water through the coil are in the _____ direction to each other.

    (a) Water flow   (c) Same

    (b) Parallel   (d) Opposite

71. On the installation of condensate drains or of well water disposal, it is not permissable to run these pipe lines to a direct connection to a city sewer.

    (a) True   (b) False

72. The actual vertical distance through which a condenser or evaporator water pump must lift water is called the:

    (a) Static head   (c) Friction head

    (b) Pressure head   (d) Vertical head

73. On a pneumatic control system, the main line pressure should be how many psig?

    (a) 25 psig   (c) 15 psig

    (b) 5 psig   (d) 60 psig

74. A pneumatic control valve which requires air pressure on the bellows or diaphram to close the valve:

    (a) Diverting valve   (c) Reverse-acting valve

    (b) Direct-acting valve (d) Normally closed valve

75. What is the unit of current or rate of flow of electricity called:

    (a) Power

    (c) Volt

    (b) Watt

    (d) Ampere

76. Which of these applications is not a mechanical refrigeration system:

    (a) Cascade system

    (b) Evaporative cooling

    (c) Absorption refrigeration

    (d) Centrifugal refrigeration

77. Refrigerant 12 and 22 may be interchanged in a system because:

    (a) They will not react chemically with each other

    (b) They are both halocarbons

    (c) All Freons have an interchangeable affinity

    (d) They should never be interchanged

78. A compressor is short cycling, which of the following would not be the cause?

    (a) Low pressure controller differential set to close

    (b) High pressure controller set to close

    (c) A leaky compressor valve

    (d) Improper thermostat operation

79. When checking for leaks with a halide torch, the flame color will be _____ when _____ % of halocarbon is in the air.

    (a) Bright green -- .01%

    (b) Bright green -- 5.0%

    (c) Bright red   -- .01%

    (d) Yellow   -- 5.0%

80. A heat pump is a refrigeration system because_____.

    (a) It is capable of pumping heat as well as cold

    (b) It can cool below standard air temperature

    (c) The refrigerant absorbs heat at one temperature and rejects it at a higher temperature.

81. A reverse acting pneumatic humidistat increases the air pressure to a controlled device when the humidity:

    (a) Increases   (c) Remains static

    (b) Decreases   (d) Mixes

82. For a given valve opening, the flow through the valve will increase as the pressure drop _____.

    (a) Increases   (c) Remains constant

    (b) Decreases   (d) Stops

83. The bimetal strip in a thermostat is a _____.

    (a) The anticipator strip

    (b) The differential strip

    (c) The contactor strip

    (d) None of these

84. On a pneumatic control system the thermostat is connected to the valve or damper motor by a small diameter tubing usually:

    (a) 1/4" O.D.   (c) 3/8" I.D.

    (b) 3/8" O.D.   (d) 1/4" I.D.

85. For a heating application using a normally open valve the controller is:

    (a) Submaster control

    (b) Indirect acting run thermostat

    (c) Direct acting run thermostat

    (d) Manual switch control

86. On a refrigeration system the compressor may be used for evacuating the system.

    (a) True   (b) False

87. All drive belts on refrigeration, air conditioning or ventilating equipment should be checked for proper tension and alignment. If each belt can be depressed about _____ with normal pressure from the thumb, the tension is about right.

    (a) 1/8" to 1/4"   (c) 3/4" to 1"

    (b) 1 pound   (d) 2" to 3"

88. A valve which has one inlet connection and two outlet connections and two separate discs and seats is known as:

    (a) Modulating plug valve

    (b) Diverting valve

    (c) 3-way mix valve

    (d) Butterfly valve

89. The term applied to a device used to accumulate the effect of two or more thermostats to operate a single device is:

    (a) Cumulator   (c) Compensator

    (b) Accumulator   (d) Coordinator

90. On a single inlet blower fan, the driving side is on what side in relation to the inlet of the fan:

    (a) Neither side   (c) Opposite side

    (b) Either side   (d) Same side

91. One of the three fundamental blower laws states that the power varies at the _____ of the speed.

    (a) Rate   (c) Square

    (b) Cube   (d) Circumference

92. What is the size of grille or register to pass 800 cfm at 600 foot velocity? Assume free area at 80%.

    (a) 133 sq in.   (c) 239 sq in.

    (b) 197 sq in.   (d) 314 sq in.

93. Most heat removed from the refrigerant in a condenser is of the _____ variety.

    (a) Latent heat   (c) Superheat

    (b) Specific heat   (d) Overheat

94. A unit of power equal to one horsepower delivered at the shaft of an engine or motor:

    (a) Electrical horsepower

    (b) Horsepower

    (c) Brake horsepower

    (d) Input power

95. According to American Standard Safety Code for Mechanical Refrigeration, what shall be provided on all systems containing more than 20 pounds of refrigerant:

    (a) Firestats

    (b) Pressure limiting devices

    (c) Receivers

    (d) Fire extinguishers

96. On a water cooled condenser operating off a cooling tower, approximately how many gallons are required per ton of refrigerant to be circulated through condenser:

    (a) 3 gallon per minute   (c) 4 gallon per minute

    (b) 2 gallon per minute

97. The unit of pressure commonly used in air ducts is measured in:

    (a) Inches of water   (c) Velocity

    (b) Inches of air   (d) Inches of mercury

98. Fans that have flow within the wheel that is substantially parallel to the shaft are known as:

    (a) Forward curved fans   (c) Centrifugal fans

    (b) Radial fans   (d) Axial fans

99. How may the thermostat be adjusted to operate at a cooler setting :

    (a) Adjust the differential

    (b) Adjust the range

    (c) Mount it in a draft

    (d) Mount closer to ceiling

100. On a refrigeration system when an air condenser or an evaporative condenser is installed in a location where it can become much warmer than the compressor, what should be installed in the hot gas line just ahead of the condenser shut-off valve:

    (a) Pressure relief   (c) Dryer

    (b) Check valve   (d) Equalizing line

SERVICE AND MAINTENANCE

ANSWER SHEET

1 - D	26 - A	51 - C	76 - B
2 - B	27 - C	52 - A	77 - D
3 - D	28 - D	53 - B	78 - C
4 - C	29 - C	54 - A	79 - A
5 - C	30 - C	55 - D	80 - C
6 - A	31 - D	56 - A	81 - B
7 - C	32 - B	57 - B	82 - A
8 - B	33 - A	58 - B	83 - D
9 - D	34 - A	59 - B	84 - A
10 - A	35 - A	60 - C	85 - C
11 - C	36 - A	61 - B	86 - B
12 - C	37 - D	62 - B	87 - C
13 - D	38 - A	63 - D	88 - B
14 - D	39 - A	64 - B	89 - A
15 - B	40 - A	65 - A	90 - C
16 - A	41 - C	66 - A	91 - B
17 - B	42 - B	67 - A	92 - C
18 - A	43 - C	68 - B	93 - A
19 - B	44 - A	69 - D	94 - C
20 - A	45 - C	70 - D	95 - B
21 - D	46 - B	71 - A	96 - A
22 - B	47 - B	72 - A	97 - A
23 - A	48 - D	73 - C	98 - D
24 - B	49 - C	74 - B	99 - B
25 - A	50 - A	75 - D	100 - B

CLOSED BOOK

4 HOUR EXAMINATION

1. Ohm's law states _____.

    (a) $I = \dfrac{R}{E}$ (c) $R = \dfrac{I}{E}$

    (b) $R = ER$ (d) $I = \dfrac{E}{R}$

2. An ampmeter reads 110 volts and 12 amperes, how many kilowatts are being consumed?

    (a) 1.32   (c) 12,000

    (b) 3.21   (d) 120

3. Low suction pressure may be caused by _____.

    (a) Dirty filters

    (b) Restriction in liquid line

    (c) Shortage of refrigerant

    (d) Any of these

4. Low discharge pressure may be caused by _____.

    (a) Dirty filters

    (b) Restriction in liquid line

    (c) Shortage of refrigerant

    (d) None of these

5. An A.C. single phase motor, capator start, will start up but heats rapidly. The cause may be _____.

    (a) Centrifugal starting switch not opening

    (b) Grounded armature winding

    (c) Open starting winding

    (d) Motor overloaded

6. An A.C. single phase motor, split phase, will not start. The cause may be _____.

    (a) Defective capacitor

    (b) High mica between commutar bars

    (c) Short circuit in the armature windings

    (d) Open circuit in motor winding

7. On an F-12 Refrigeration System, if the suction pressure at the compressor is 35 psi with a 2 psi suction line loss and the suction line temperature taken at point TX valve bulb is fastened, is 51°, what is your Superheat Reading:

    (a) 13°F   (c) 11°F

    (b) 14°F   (d) 9°F

8. Which of the following electrical wires will carry the most current:

    (a) 14   (c) 12

    (b) 10   (d) 24

9. An electrical network which has 120 volts to a neutral from all legs:

    (a) Delta   (c) Polyphase

    (b) Star   (d) 2 phase

10. What type of threads are used on flare fittings:

    (a) National fine   (c) U.S.A.

    (b) National course   (d) Standard pipe threads

11. What fitting should be used to connect 1/4" O.D. copper tubing to a 1/4" internal pipe thread opening in the compressor body:

    (a) A union   (c) A tee fitting

    (b) A half-union   (d) A street ell

12. When using a double element time delay fuse with a 20 amp. motor the maximum fuse size should be:

    (a) 60 amps.   (c) 30 amps.

    (b) 20 amps.   (d) 25 amps.

13. On a motor, a running capacitor is used to:

    (a) Increase torque   (c) Increase power factor

    (b) Increase horsepower   (d) Increase amperage

14. If, upon letting an R-12 refrigeration compressor stand idle long enough for the entire system to cool down to the temperature of the surrounding air of 90°F and the reading of the head pressure gauge is 115 lb., what is this an indication of:

    (a) Overcharge of gas   (c) Air in system

    (b) Undercharge of gas   (d) Vacuum gauge

15. What is used to determine the amount of moisture remaining in a refrigeration system:

    (a) Vacuum pump   (c) Wet bulb indicator

    (b) Vacuum dehydration   (d) Vacuum gauge
        indicator

16. In the discussion of the refrigeration cycle, Compression Ratio is expressed in the following equation thus:

    (a) Compression Ratio $= \dfrac{\text{Discharge pressure PSI absolute}}{\text{Suction pressure PSI absolute}}$

    (b) Compression Ratio $= \dfrac{\text{Suction pressure PSI absolute}}{\text{Discharge pressure PSI absolute}}$

    (c) Compression Ratio $= \dfrac{\text{Discharge temperature absolute}}{\text{Suction temperature absolute}}$

    (d) Compression Ratio $= \dfrac{\text{Suction temperature absolute}}{\text{Discharge temperature absolute}}$

17. An electrical condenser used for power factor correction is known as a :

   (a) Transformer   (c) Capacitor

   (b) Starter   (d) Relay

18. A ton of refrigeration is a unit of refrigeration capacity corresponding to the removal of how many BTU per day:

   (a) 144,000 Btu's   (c) 288,000 Btu's

   (b) 200,000 Btu's   (d) 12,000 Btu's

19. A thermal device the opens its contacts when the electric current through a heater coil exceeds the specified value for a given time is called:

   (a) Fuse   (c) Control relay

   (b) Relief relay   (d) Thermal overload relay

20. What should be first adjustment when adjusting a control which has a cut in differential adjustment:

   (a) Differential

   (b) Gauge for cut-in

   (c) Either differential or range

   (d) Range for cut-out

21. The gas resulting from the instantaneous evaporation of refrigerant in a pressure-reducing device to cool the refrigerant to the evaporation temperature obtained at the reduced pressure:

   (a) Superheat gas   (c) Inert gas

   (b) Flash gas   (d) Non-condensible gas

22. If three unmarked gas cylinders R-12, R-22 and R-500 respectively are all about one-half full and have been stored in the same room for several days at a temperature of 80°F which cylinder contains the R-22:

    (a) There is no way to tell if not marked

    (b) The cylinder that gives a gauge reading of 103#

    (c) The cylinder that gives a gauge reading of 84#

    (d) The cylinder that gives a gauge reading of 145#

23. Mercury is used in thermometers because it:

    (a) Expands uniformly

    (b) Does not freeze until it reaches 38.2°F

    (c) Does not boil until it reaches 674.6°F

    (d) Does all of the above

24. What will the reading of Wet and Dry Hygrometers be if exposed to air that is completely saturated:

    (a) They will both read alike

    (b) The Wet Bulb will be lower

    (c) The Dry Bulb will be lower

    (d) 90% relative humidity

25. What type of electric motor has a set of field coils and a rotating armature:

    (a) An induction motor

    (b) A capacitor split phase motor

    (c) A repulsion start induction run motor

    (d) Dual winding repulsion induction motor

26. When a dry and wet bulb thermometer are held in a dry air stream the wet bulb will always read lower because _____.

    (a) A wet bulb thermometer is calibrated lower

    (b) The wet bulb thermometer is cooled by evaporization

    (c) A wet bulb thermometer is always mercury

    (d) It won't read lower

27. When an air sample precipitates moisture, it indicates _____.

    (a) The leaving air is too warm

    (b) The leaving air is too cold

    (c) The air is not circulating across the coil

    (d) The saturation temperature has been passed

28. A squirrel gage fan is another name for a _____.

    (a) Backward inclined fan

    (b) Forward inclined fan

    (c) A radial flow fan

    (d) An exhauster fan

29. What is common to all filters?

    (a) They must all be coated

    (b) They are all constructed of fiber glass

    (c) They are all 100% fireproof

    (d) They all have a pressure drop

30. A nominal one ton 115 volt window unit is located 50 ft from current source, the electric wire size should be:

    (a) #12   (c) #16

    (b) #10   (d) #8

31. A properly operating 1 ton window unit without a dehumidifier will _____.

    (a) Not dehumidify the air

    (b) Dehumidify the air

    (c) Dehumidify the air only when it is saturated

    (d) Dehumidify the air when it reaches saturation

32. A window unit which rumbles when the compressor stops may _____.

    (a) Not have compressor shipping bolts

    (b) Have a ruptured capillary

    (c) Not be mounted level

    (d) Be polyphased

33. Air leaving the condenser should always _____.

    (a) Have an unrestricted path to outdoors

    (b) Be pumped out through the exhaust

    (c) Be filtered

    (d) Have an unrestricted path to the room

34. A room cooler in normal operation will remove _____ heat.

    (a) Sensible   (c) Latent and specific

    (b) Latent   (d) Sensible and latent

35. The bimetal strip in a thermostat is _____.

    (a) The anticipator strip

    (b) The differential strip

    (c) The contractor strip

    (d) None of these

36. A heat pump system may meter the refrigerant by using _____.

    (a) Only a capillary tube

    (b) Only a CP tube

    (c) Only a TX valve

    (d) A capillary tube or TX valve

37. In heat pump terminology, CP is _____.

    (a) A kind of tube

    (b) Condensing pressure

    (c) Coefficient of performance

    (d) Constant pumping system

38. An auxiliary heat source when used with a heat pump is generally _____.

    (a) A condenser reheat   (c) Electrical heating

    (b) A booster pump   (d) Heat of compression

39. If a heat pump is designed with a ground coil heat source it is pumping _____through the coil.

    (a) H_2O (c) CCL_3F

    (b) CO_2 (d) $CHCLF_2$

40. A heat pump usually pumps _____ through the evaporator coil.

    (a) H_2O (c) CCL_2F

    (b) CO_2 (d) $CHCLF_2$

41. An R-12 system will have a _____ compressor displacement/ton cu. ft./min than an R-22 system.

    (a) Greater   (c) Equal

    (b) Lesser   (d) It depends on the Btu input

42. A heat-pump is a refrigeration system because _____.

    (a) It is capable of pumping cold as well as heat

    (b) It can cool below standard air temperature

    (c) The refrigerant absorbs heat at one temperature and rejects it at a higher temperature

    (d) It has a four-way refrigeration valve

43. Refrigerant 12 and 22 may be interchanged in a system because _____.

    (a) They will not react chemically with each other

    (b) They are both halocarbons

    (c) All freons have an interchangeable affinity

    (d) They should never be interchanged

44. When checking for leaks with a halide torch the flame color will be _____ when _____% of halocarbon is in the air.

    (a) Bright green - .01%

    (b) Bright green - 5.0%

    (c) Bright red - .01%

    (d) Yellow   - 5.0%

45. A compressor is short cycling, which of the following would *not* be the cause?

  (a) Low pressure controller differential set too close

  (b) High pressure controller set too close

  (c) A leaky compressor valve

  (d) Improper thermostat operation

46. What alloys shall not be used in contact with any freon refrigerant:

  (a) Aluminum   (c) Magnesium

  (b) Zinc   (d) Copper

47. A unit of power equal to one-horsepower delivered at the shaft of an engine or motor is commonly known as:

  (a) Amperage   (c) Voltage

  (b) Brake horsepower (d) Horsepower

48. What is the range of a thermostat:

  (a) The open and close settings

  (b) The temperature difference

  (c) The capacity of the thermostat

  (d) The variety of the models available

49. Heat being a form of energy cannot be:

  (a) Measured   (c) Transferred

  (b) Felt   (d) Destroyed

50. Removal of heat from compressed gas between compression stages is known as:

  (a) Recooling   (c) Subcooling

  (b) Intercooling   (d) Upcooling

ROOM AIR CONDITIONING

ANSWER SHEET

1 - D	26 - B
2 - A	27 - D
3 - D	28 - C
4 - C	29 - D
5 - A	30 - A
6 - D	31 - B
7 - C	32 - C
8 - B	33 - A
9 - B	34 - D
10 - A	35 - D
11 - B	36 - D
12 - D	37 - C
13 - C	38 - C
14 - C	39 - A
15 - B	40 - D
16 - A	41 - A
17 - C	42 - C
18 - C	43 - D
19 - D	44 - A
20 - D	45 - C
21 - B	46 - C
22 - D	47 - B
23 - D	48 - A
24 - A	49 - D
25 - A	50 - B

CLOSED BOOK

4 HOUR EXAMINATION

1. The value of "K" for any insulation is a _____ :

    (a) Conductivity factor

    (b) Transmission factor

    (c) Resistance factor

    (d) Absorption factor

2. "K" value is expressed as:

    (a) Btu per sq ft per hour per degree F from air to air

    (b) Btu per sq ft per hour per degree F from surface to surface

    (c) Btu per sq ft per hour per degree per inch thickness

    (d) None of the above

3. A typical frame wall is constructed of the following materials, where the resistance factor is known:

1.	Outside air @ 15 mph wind	R	0.17
2.	Wood siding	R	0.85
3.	Insulation sheathing	R	2.06
4.	Air space	R	0.97
5.	3/8" gyp lath & plaster	R	0.41
6.	Inside air - still	R	0.68

The Btu of the "U" factor is _____ .

    (a) 5.14   (c) 0.27

    (b) 4.27   (d) 0.19

4. The thermal conductances of individual materials are:

    (a) Additive to each other

    (b) Multiplied by each other

    (c) Have a logarithmic progression

    (d) None of the above

5. It is required that a roof construction have a total "U" value of 0.13 Btu. The construction is as follows:

1.	Outside air @ 15 mph	R	0.17
2.	Built up roofing	R	0.33
3.	2" poured gypsum deck	R	1.20
4.	1/2" gyp form board	R	0.45
5.	Inside air - still	R	0.61

    It is necessary to insulate the roof with a material of _____ resistance to meet the "U" 0.13:

    (a) 2.76   (c) 4.93

    (b) 0.76   (d) 0.93

6. Given the following conditions:

    1. Pipe temperature 0° F.
    2. Pipe size 3"
    3. Ambient air 90° F, 80% R.H.
    4. Air-film coefficient 1.65
    5. K factor of insulation 0.24

    The required thickness of the insulation to prevent condensation would be_____:

    (a) 1 inch   (c) 2 inches

    (b) 1-1/2 inches   (d) 2-1/2 inches

7. The recommended method of determining required insulation thickness is based on _____ and prevention of condensation for low temperature installations.

    (a) Temperature differential

    (b) 90% efficiency

    (c) Permeability

    (d) Economic analysis

8. The most important characteristic of materials for underground insulation is _____:

    (a) Lack of moisture absorption

    (b) Low heat conductivity

    (c) Ability to withstand physical shock

    (d) Ability to withstand thermal shock

9. All low-temperature insulations (below ambient) should have vapor barrier covering joint sealants and adhesives with a maximum value of _____ perm:

    (a) 0.05   (c) 0.15

    (b) 0.10   (d) 1.00

10. According to the recommended method in Question #7, when insulating heated piping, the fittings, valves and flanges _____:

    (a) Require equal protection

    (b) Do not require equal protection

    (c) Should be trowelled with foamed-plastic

    (d) Must have a protective jacket

11. Given a below ground insulation job in the temperature range of 300° F which of the following insulations should be avoided because of lack of strength and tendency to deteriorate through moisture:

    (a) Calcium silicate

    (b) Preformed glass fiber

    (c) Preformed cellular glass

    (d) 85% magnesia

12. An above ground outside chilled water line in the temperature range of 5° C should be insulated with:

    (a) Calcium silicate

    (b) Diatomaceous silicate

    (c) Magnesium carbonate & asbestos

    (d) Cellular glass

13. Given a loose fill insulation job for pipe lines in an underground conduit you would compact to minimum density of _____ lbs. per cubic foot:

    (a) 15   (c) 5

    (b) 7-1/2   (d) 3-1/2

14. According to ASTM & MIL specifications which of the following is acceptable for maximum temperatures to 1900° F in blocks or boards:

    (a) Diatomaceous silica

    (b) 85% magnesium

    (c) Calcium silicate

    (d) Foamed cellular glass

15. When using Armaflex to insulate screwed pipe, the fitting covers should _____:

    (a) Be mitered and butted to the pipe insulation

    (b) Be slitted and wrapped

    (c) Be undercut and buttered with Armstrong 520

    (d) Overlap pipe insulation one inch

16. A single 24"x18" supply duct running 40 ft through a
    95° F D.B. 75° F W.B. unconditioned space and deliv-
    ering air to one single ceiling diffuser is connected
    to a coil with leaving air 58.4 F D.B. 56.1 F W.B.
    what would be the heat gain to the duct after it had
    been insulated with 1" Ultralite #200 (K.24):

    (a)   246 Btu/hr   (c)   2460 Btu/hr

    (b)   1245 Btu/hr   (d)   12460 Btu/hr

17. A duct liner is generally _____:

    (a)   Carefully cut with individual pieces to fit

    (b)   Applies to the flat metal with adhesive and
          fabricated in a brake

    (c)   Slipped in fitted sections and fastened

    (d)   Reverse snapped

18. If you were insulating a brewery with "Fiberglass PF"
    board on the walls, which of the following would *not*
    be true:

    (a)   You would have a vapor seal on the warm side
          only

    (b)   You would have a vapor porous material on the
          cold side

    (c)   You would have a multiple vapor barrier
          solution

    (d)   You would apply the boards by pressing them
          into place between wood joists 4" thick

19. A 25° F cold storage room requires a vapor barrier
    having permeance not exceeding _____ perms:

    (a)   0.2   (c)   0.01

    (b)   0.1   (d)   0.001

20. A federally inspected meat processing plant *must* have:

    (a) 6' wide door openings

    (b) Metal clad doors

    (c) Tile floors throughout

    (d) Minimum of 4" insulation

21. If a spec called for "Laykold" insulation adhesive for a cold storage job using fiberglass, which of the following would *not* be true:

    (a) The consistency of "Laykold" permits spraying on

    (b) The consistency of "Laykold" permits brushing on

    (c) It would not be suitable for use at elevated temperatures

    (d) It would require 30 to 60 minutes of setting time before being covered

22. If you were doing a cold storage room with Fiberglas AE-F board, you would _____:

    (a) Lay it with broken joints on the floor only

    (b) Lay it in hot asphalt on all the walls

    (c) Cover with a vapor seal and trowel a finish over it

    (d) Use the metal clip method of application

23. A "Jamison Flexidor" is _____:

    (a) An overhead type door

    (b) Constructed of neoprene

    (c) An electric sliding door

    (d) Used exclusively for banana rooms

24. A cold storage door must have the following safety
    feature _____:

    (a) A flashing red light when anyone is inside

    (b) An alarm bell which can be operated from
        inside of the storage room

    (c) A counter balance closure

    (d) A fastener release with provision to open the
        door from inside even though it is padlocked
        on the outside

25. Which of the following is uncommon to the group by
    material:

    (a) K & M Hytemp   (c) Carey Hi-temp

    (b) JM Superex   (d) Paco Caltemp

26. Which of the following is uncommon to the group by
    temperature range:

    (a) OC Kaylo   (c) Pabco Prasco

    (b) JM Thermobestos   (d) K&M Kaytherm

27. Which of the following is uncommon to the group by
    material and temperature range:

    (a) PC Foamglas   (c) JM Microlok

    (b) LOF Superglas   (d) GB Snap-on

28. Which of the following is uncommon to the group by
    temperature range:

    (a) Dow Styrofoam   (c) JM Zerolite

    (b) JM Sil-0-Cel   (d) BH Coldboard

29. Which of the following is uncommon to the group by
    material:

    (a) Foamglas   (c) Armalite

    (b) Styrofoam   (d) Unicrest

30. On pipe insulation bands are generally used to:

    (a) Hold the insulation in place

    (b) Identify the insulation

    (c) Give the job a good appearance

    (d) Seal the joints

31. Foamed cellular glass pipe insulation comes in standard lengths of _____:

    (a) 24" and 36"   (c) 18" and 36"

    (b) 18" and 24"   (d) 36"

32 The standard length for most pipe coverings is:

    (a) 24" and 36"   (c) 18" and 36"

    (b) 18" and 24"   (d) 36"

33. Thermal insulation is *not* used to _____:

    (a) Retard heat flow through a surface

    (b) Control the temperature inside of a vessel

    (c) Prevent condensation

    (d) Control the surface temperature of a vessel

34. For purposes of taking-off an estimate, the saddle brackets for insulated pipe are generally considered part of the _____ responsibility of specs:

    (a) Piping contractor's

    (b) Sheet metal contractor's

    (c) Insulation contractor's

    (d) Iron worker's

35. When taking off an estimate it is standard to consider which of the following as *incorrect:*

    (a) Weld valve = one fitting

    (b) Weld ell   = one fitting

    (c) Weld tee   = three fittings

    (d) Flanged valve = one fitting and 2 pr. flanges

36. For estimating insulation pipe covering in equivalent length of pipe _____ is correct procedure:

    (a) Pair of flanges = 2 ft straight pipe

    (b) Pair of flanges = 3 ft straight pipe

    (c) Pair of flanges = 2 x length

    (c) Couplings   = 2 x length

37. The area of standard dish and flat head vessel shall be _____:

    (a) The square of the inside diameter of the insulation on the shell body

    (b) The square of the outside diameter of the shell body before insulation

    (c) The square of the outside diameter of the insulation on the shell body

    (d) The length x the circumference to the inside of the insulation

38. Which of the following coatings is non-vapor barrier:

    (a) Eagle-Pitcher   Insulseal

    (b) Carey   Insulation seal

    (c) Foster   60-26

    (d) Insulmastic   4010

39. All of the following coatings have a covering capacity of approximately 150 sq ft per gallon, except _____ which will only cover approximately 1/3 as much as the others:

    (a) Asphalt varnish   (c) Aluminum paint

    (b) Asphalt primer   (d) Asphalt emulsion

40. All of the following coatings have a covering capacity of approximately 100 sq ft per gallon, except _____ which will cover about twice as much surface per gallon:

    (a) Asphalt paint

    (b) Foster #30-76 white insulation coating

    (c) Lagging adhesive

    (d) Glue sizing 2 lb/gallon

41. The square feet of surface per lineal foot of 2-1/2 inch, 85% mag. insulated 8 in. steel pipe is _____:

    (a) 3.60   (c) 2.24

    (b) 3.08   (d) 2.09

42. The square feet of surface per lineal foot of 1-1/2 inch, 85% mag. insulated 14" steel pipe is _____:

    (a) 2.95   (c) 4.55

    (b) 3.67   (d) 5.40

43. When estimating canvas or weather-proof felt, the total sq ft in the take-off should be multiplied by _____ to allow for lapping, folding, and waste:

    (a) 105%   (c) 150%

    (b) 120%   (d) 2

44. Approximately _____ labor time should be allowed to cover 100 ft of 14 inch insulated pipe with roofing jacket:

    (a) 2 hours

    (b) 5 hours

    (c) 10 hours

    (d) 14 hours

45. When work is done in enclosed areas of temperatures of over 90° F, labor time should be estimated at an additional _____ above regular work:

    (a) 1% per each °F

    (b) 5% per each °F

    (c) 10% per each °F

    (d) Flat 50%

46. Dual temperature pipe insulation is usually _____:

    (a) Broken joint application sealed with asbestos cement

    (b) Segmental form with a coat of lagging cement

    (c) Split longitudinal, buttered and wired

    (d) Single layer, stapled and vapor barrier sealed

47. Foamed cellular glass pipe covering 1-1/2" thickness is generally applied in segmental form on pipe sizes from _____ inches and over:

    (a) 10

    (b) 12

    (c) 14

    (d) 18

48. 1/2" thick fiberglas with light weight foil laminate jacket is acceptable pipe insulation for _____:

    (a) Low temperature range, minus 30° F to zero

    (b) Cryogenics

    (c) Weather exposed lines

    (d) Medium temperature range 50° F to 60° F

49. Metal mesh blanket insulation for tanks, ovens , etc., up to 1000° F generally comes in widths of _____ inches:

    (a) 24   (c) 36

    (b) 30   (d) 48

50. Good quality pour-in-place urethane insulation is most likely to have a K factor (75° F 2 lb/cu ft density) of _____:

    (a) 0.0155   (c) 0.25

    (b) 0.155   (d) 0.3

MASTER OF INSULATION

ANSWER SHEET

1	-	A		
2	-	C		
3	-	D		
4	-	D		
5	-	C		

<table>
<tr><td>1 - A</td><td>26 - C</td></tr>
<tr><td>2 - C</td><td>27 - A</td></tr>
<tr><td>3 - D</td><td>28 - B</td></tr>
<tr><td>4 - D</td><td>29 - A</td></tr>
<tr><td>5 - C</td><td>30 - C</td></tr>
<tr><td>6 - B</td><td>31 - B</td></tr>
<tr><td>7 - D</td><td>32 - D</td></tr>
<tr><td>8 - A</td><td>33 - B</td></tr>
<tr><td>9 - B</td><td>34 - A</td></tr>
<tr><td>10 - A</td><td>35 - C</td></tr>
<tr><td>11 - D</td><td>36 - B</td></tr>
<tr><td>12 - D</td><td>37 - C</td></tr>
<tr><td>13 - B</td><td>38 - A</td></tr>
<tr><td>14 - A</td><td>39 - D</td></tr>
<tr><td>15 - D</td><td>40 - D</td></tr>
<tr><td>16 - C</td><td>41 - A</td></tr>
<tr><td>17 - B</td><td>42 - C</td></tr>
<tr><td>18 - C</td><td>43 - B</td></tr>
<tr><td>19 - C</td><td>44 - C</td></tr>
<tr><td>20 - B</td><td>45 - A</td></tr>
<tr><td>21 - D</td><td>46 - D</td></tr>
<tr><td>22 - A</td><td>47 - C</td></tr>
<tr><td>23 - B</td><td>48 - D</td></tr>
<tr><td>24 - D</td><td>49 - A</td></tr>
<tr><td>25 - D</td><td>50 - B</td></tr>
</table>

OPEN BOOK

4 HOUR EXAMINATION

VALUE:

| 1 through 78 | 58 points | 86 | 10 points |
| 79 through 85 | 14 points | 87 through 89 | 18 points |

1. By increasing a vent stack from 4" to 8" capacity will increase:

    (a) 2-1/2 times   (c) 2 times

    (b) 4 times   (d) 3 times

2. A smoke stack has a circumference of 31.416" what is the diameter:

    (a) 11"   (c) 10"

    (b) 16-1/2"   (d) 12"

3. One of the following would be desirable velocity for a low velocity system:

    (a) 500 fpm   (c) 1500 fpm

    (b) 1000 fpm   (d) 2000 fpm

4. Per 100 ft of duct what would be a desirable friction loss:

    (a) .08   (c) .02

    (b) .12   (d) .05

5. Duct design criteria is not based on the following:

    (a) Friction loss   (c) Btu capacity

    (b) Velocity   (d) Static pressure

6. The weight of 26 ga. galvanized sheet per square foot is:

    (a) .906 lbs   (c) 1.010 lbs

    (b) .708 lbs   (d) .560 lbs

7. Inspection hand holes on range hood exhaust shall be installed:

    (a) Top of range hood   (c) At filter section

    (b) Side of duct   (d) Bottom side of
                                   horizontal duct

8. Uninsulated breeching for solid fuel burning furnace may not approach combustible material less than:

    (a) 18"   (c) 24"

    (b) 10"   (d) 6"

9. Vertical risers for range hood exhaust passing through floors shall have the vertical duct enclosed in a shaft of:

    (a) 4" hollow tile or masonry

    (b) Lath and plaster

    (c) Wrapped with blanket fiberglass

    (d) Enclosed in fire resistant insulation

10. Gas filled furnace has 6" flue stack, what other exhaust may be connected to this stack:

    (a) Toilet vent   (c) Laundry vent

    (b) Kitchen range hood   (d) None other

11. In connecting duct work to a unit with two fan outlets, a common practice is to use:

    (a) Pittsburgh lock   (c) Pair of pants

    (b) Mixing box   (d) Splitter damper

12. Triangulation is a term used in:

    (a) Duct layout   (c) Duct sizing

    (b) Duct insulation   (d) None of the above

13. A standing seam used as a cross seam for large duct is used to:

    (a) When flat surface is required

    (b) Eliminate the need for angle reinforcement

    (c) Is used for light gage duct

    (d) It is simple to make

14. Describe a duct transition:

    (a) Fitting changing duct size or shape

    (b) Fitting for spitter damper

    (c) Fitting for branch take off

    (d) Fitting for fire damper

15. A true grooved seam allowance is:

    (a) 1-1/2 times the width of seam

    (b) 2 times the width of seam

    (c) 3 times the width of seam

    (d) 2-1/2 times the width of seam

16. Rectangular duct is 32"x18" the equivalent in round duct would be:

    (a) 20"   (c) 30"

    (b) 26"   (d) 22"

17. The approximate weight per lineal foot of angle iron 1"x1"x1/8" is:

    (a) .84#   (c) 1.02#

    (b) .44#   (d) .96#

18. In taking off duct to determine weight, elbows are always measured:

    (a) On center line   (c) On the radius

    (b) Twice   (d) On the perimeter

19. Where permitted metal smoke stacks from 8" to 12" diameter are made of:

    (a) 12 gal.   (c) 18 gal.

    (b) 16 gal.   (d) 20 gal.

20. Metal smoke stacks within 25 ft of any building must extend not less than:

    (a) 8 ft   (c) 2 ft

    (b) 4 ft   (d) 6 ft

21. On gas fired warm air furnace the closet is never used as a return air plenum because:

    (a) Mixing of combustion gas would cause sickness or death

    (b) Cooking smell would circulate

    (c) Use up heated air

    (d) Heated air would rise to attic

22. Metal smoke stacks may be used in buildings of the following occupancy:

    (a) Type A   (c) Type H

    (b) Type I   (d) None of the above

23. Flues for oil fired warm air furnace shall be:

    (a) Type A   (c) Type C

    (b) Type B   (d) Type D

24. Ducts conveying exhaust from hoods over commercial kitchen range should have a conveying minimum velocity of:

    (a) 1000 fpm   (c) 2000 fpm

    (b) 1500 fpm   (d) 500 fpm

25. Air velocity across the face of range hoods shall be:

    (a) 100 fpm   (c) 25 fpm

    (b) 50 fpm   (d) 75 fpm

26. Hoods for kitchen ranges should have a clearance from combustible material of:

    (a) 12"   (c) 4"

    (b) 6"   (d) 2"

27. Duct connection to range hoods shall be made at the:

    (a) Side of hood   (c) End of hood

    (b) Top of hood   (d) Bottom of hood

28. Range hoods shall be hung within the following distance from the floor:

    (a) 6' to 7'   (c) 5' to 6'

    (b) 7' to 8'   (d) Over 8'

29. Warehouse area power roof ventilation must have:

    (a) Backdraft damper   (c) Firestat

    (b) Fusestat   (d) Birdscreen

30. Mechanical ventilation for paint spray booths must have:

    (a) Explosion proof motor

    (b) Two speed motor

    (c) Back draft damper

    (d) Fusestat

31. What instrument shall be used to check motor performance on a mechanical ventilation system:

    (a) Pressure gage   (c) Pitot tube

    (b) Ampmeter   (d) Velometer

32. To check the rpm of a belt driven fan what instrument would be used:

    (a) Ampmeter   (c) Velometer

    (b) Tachometer   (d) Manometer

33. To check air volume at a side wall supply grill what instrument may be used:

    (a) Tachometer   (c) Ampmeter

    (b) Velometer   (d) Mercury gage

34. Copper gravel-stops for gravel roofs must have a minimum weight of:

    (a) 16 oz.   (c) 32 oz.

    (b) 24 oz.   (d) 12 oz.

35. Lay-out development of a gutter miter is called:

    (a) Triangulation   (c) Radial

    (b) Parallel lines   (d) Progression

36. Lay-out development for a funnel pattern is:

    (a) Radial   (c) Progression

    (b) Triangulation   (d) Parallel lines

37. Galvanized sheet metal is made by applying a protective coating of:

    (a) Lead   (c) Tin

    (b) Zinc   (d) Nickel

38. Copper may be annealed by:

    (a) Heat and cool slowly

    (b) Heat and-quench

    (c) Apply electricity

    (d) None of the above

39. Duct size is 18" diameter, friction loss .10 cfm 2000, what is the fpm velocity:

    (a) 800   (c) 600

    (b) 1150   (d) 1500

40. Duct velocity of 750 fpm, friction loss .0475, cfm 1300, what is the diameter of round duct:

    (a) 18"   (c) 10"

    (b) 20"   (d) 25"

41. Duct is 10" diameter, fpm velocity is 680, friction loss .08, what is the cfm:

    (a) 360   (c) 760

    (b) 560   (d) 860

42. Duct is 14" diameter, fpm velocity is 960, with 1000 cfm, what is the friction loss:

    (a) 0.06   (c) 0.10

    (b) 0.08   (d) 0.15

43. Duct is 6" diameter, velocity is 500, friction loss is .08, what is the cfm:

    (a) 200   (c) 400

    (b) 300   (d) 100

44. The following tool is used to cut out the cheeks of a 20 ga. elbow:

    (a) Groover   (c) Duck bills

    (b) Snips   (d) Dividers

45. When setting a seam by hand on a round pipe you would use a:

    (a) Duck bill   (c) Seams

    (b) Rivet set   (d) Punch

46. The Pittsburgh lock is used on:

    (a) Square duct

    (b) Round pipe

    (c) Galvanized metal only

    (d) Expanded metal

47. Which of the following would not likly be found in a sheet metal man's tool box:

    (a) Snips

    (b) Vise grips

    (c) Dividers

    (d) Torque wrench

48. The width of a steel rule 36" or 48" long used on a bench for layout is:

    (a) 1-1/4" wide

    (b) 2" wide

    (c) 3/4" wide

    (d) None of the above

49. The width of a 6 foot folding rule, wood or metal is:

    (a) 3/8"

    (b) 1/2"

    (c) 5/8"

    (d) 3/4"

50. When hand seaming metal one must consider:

    (a) Metal thickness

    (b) Paint allowance

    (c) Soldering

    (d) None of the above

51. Duct work is bent up on a:

    (a) Ductulator

    (b) Slitter

    (c) Lock former

    (d) Brake

52. Drive cleats can be made on a:

    (a) Seamer

    (b) Bar folder

    (c) Anvil or rail

    (d) None of the above

53. When drawing rivets, one would use a:

    (a) Rivet set

    (b) Whitney punch

    (c) Awl

    (d) Divider

54. A soldering iron is made of copper becuase:

   (a) Specific gravity is high

   (b) It is not affected by acid

   (c) Good conductor of heat

   (d) It is resistant to fire

55. When tinning a soldering iron you would use:

   (a) Acid   (c) 95/5 solder

   (b) Zinc   (d) Sal-A-Monac

56. Aviation snips are sharpened:

   (a) On a grinder   (c) By a special file

   (b) On a stone   (d) By tightening bolt

57. When drilling a hole for a #8 screw you would use a:

   (a) 3/16" drill bit   (c) 3/8" drill bit

   (b) 1/4" drill bit   (d) #30 drill bit

58. To spot weld copper you would have to:

   (a) Change points   (c) Use low heat

   (b) Use high heat   (d) None of the above

59. Where outside air louvers are located on the roof the recommended minimum height from roof should be:

   (a) 6 inches   (c) 16 inches

   (b) 2.5 feet   (d) 3.2 feet

60. The square root of 625 is:

   (a) 15   (c) 25

   (b) 19   (d) 29

61. Obstruction of any kind must be avoided inside high velocity ducts as they may cause:

    (a) Turning of air flow

    (b) Slowing down fan motor

    (c) Noise in air stream

    (d) Cause fan to turn backward

62. To find the area of a circle the formula is:

    (a) $\pi^2 R$ (c) $D^2 \times 8.7$

    (b) πR^2 (d) $2 \times 1/2\ D \times \pi$

63. A duct system is considered high velocity when air travels in excess of:

    (a) 12 miles per hour   (c) 800 gpm

    (b) 15,000 cfm   (d) 2,500 fpm

64. How many degrees are there in a semi-circle:

    (a) 45 degrees   (c) 180 degrees

    (b) 90 degrees   (d) 360 degrees

65. In low velocity air distribution systems the flow of air to the branch take offs is regulated by which of the following?

    (a) Amprobe   (c) Splitter damper

    (b) Volume controller   (d) Draft indicator

66. Which of the following is the nearest correct statement describing a true transition fitting:

    (a) Changing fitting from one size to another

    (b) Changing fitting from one shape to another

    (c) Changing fitting from one job to another

    (d) Changing fitting from trunk line to branch

67. Find the cubic capacity in cubic feet of a rectangular box that is 2'6" x 48" x 12':

    (a) 100 cu ft

    (c) 120 cu ft

    (b) 110 cu ft

    (d) 130 cu ft

68. The diameter of a circle 22" in circumference is:

    (a) 6-1/2 in.

    (c) 7-1/2 in.

    (b) 7 in.

    (d) 8 in.

69. In high velocity systems the supply ducts are normally limited to a maximum duct velocity of:

    (a) 4000 fpm

    (c) 1800 cfm

    (b) 2000 cfm

    (d) 5000 fpm

70. The area of a circle 16" in diameter is:

    (a) 201.06 sq in.

    (c) 102.6 sq in.

    (b) 201.60 sq in.

71. Range hoods should be hung within which of the following distances from the floor:

    (a) 5' to 6'

    (c) 6' to 7'

    (b) 8' to 9'

    (d) 9' to 10'

72. Ducts from hoods over ranges and fry kettles shall be constructed of the following gauge (or heavier) sheet metal:

    (a) 24 gauge

    (c) 20 gauge

    (b) 22 gauge

    (d) 26 gauge

73. Exhaust duct connections shall be made to range hoods at the:

    (a) Top of the hoods

    (c) End of the hoods

    (b) Bottom of the hoods

    (d) Back of the hoods

74. In ducts 61" and over on any side, the following bracing and reinforcing is recommended:

    (a) Pittsburgh drive

    (b) Reinforced pocket slip or bar "S"

    (c) Reinforced bucket roll "Y"

    (d) Pittsburgh lock

75. Hoods over kitchen cooking equipment shall be constructed of the following material:

    (a) Plastic vinyl

    (b) Marine plywood

    (c) Non-combustible material

    (d) Heat resistant material

76. Clean out openings shall be provided in every metal smoke stack at:

    (a) The base

    (b) The furnace or heat producing apparatus

    (c) Specified intervals on the duct

    (d) 6' from the floor

77. Hand holes shall be provided in range hood exhaust systems for cleaning and inspections each:

    (a) 15 feet

    (b) 20 feet

    (c) 30 feet

    (d) 40 feet

78. In high velocity systems ducts up to 24" in diameter should be:

    (a) 22 gauge

    (b) 30 gauge

    (c) 10 gauge

    (d) 60 gauge

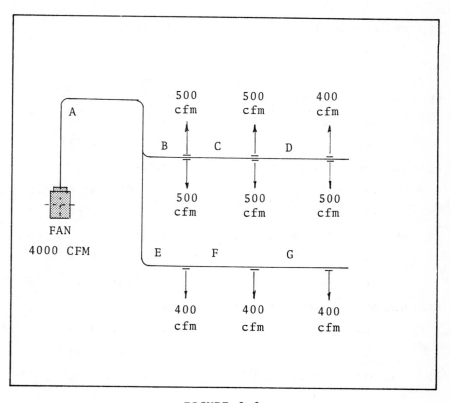

FIGURE 2.2

DUCT LAYOUT

ANSWER QUESTIONS 79-85 FROM DUCT FIGURE 2.2 USING THE EQUAL FRICTION METHOD OF DUCT SIZING.

79. Using 1600 fpm as the initial velocity at the fan, the area of duct at A is _____ sq. ft.

    (a) 2.5   (c) 3.5

    (b) 3.0   (d) 4.0

80. The area of duct at B is _____ sq. ft.

    (a) 2.5   (c) 3.0

    (b) 1.91   (d) 1.80

81. The area of duct at C is _____ sq. ft.

    (a) 1.91   (c) 1.52

    (b) 1.70   (d) 1.23

82. The area of duct at D is _____ sq. ft.

    (a) 0.52   (c) 1.28

    (b) 0.94   (d) 1.91

83. The area of duct at E is _____ sq. ft.

    (a) 0.67   (c) 1.28

    (b) 0.80   (d) 1.91

84. The area of duct at F is _____ sq. ft.

    (a) 0.52   (c) 1.28

    (b) 0.94   (d) 1.91

85. The area of duct at G is _____ sq. ft.

    (a) 0.94   (c) 0.30

    (b) 0.67   (d) 0.20

86. In the duct layout shown in Duct Figure 2.3 the total weight of galvanized iron required is _____ lbs.

    (a) 1800   (c) 3200

    (b) 2900   (d) 3900

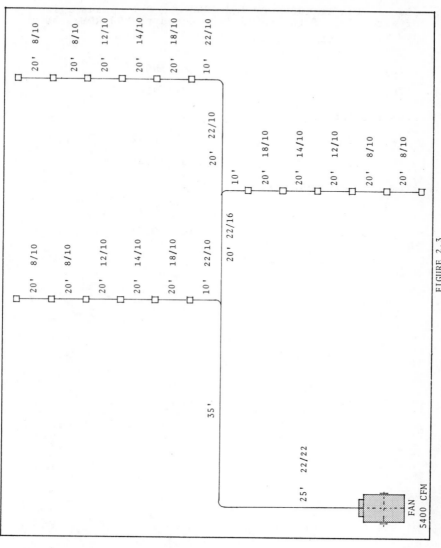

FIGURE 2.3
DUCT LAYOUT

87. Given a fan at 1000 RPM and 1200 CFM @ 1 hp and 1 in. of water Sp, if the speed is reduced to 800 RPM, the new CFM is _____.

    (a) 1,060   (c) 960

    (b) 980   (d) 920

88. From 87, the new Sp is _____.

    (a) 1.20   (c) 0.62

    (b) 0.64   (d) 0.58

89. From 87, the new hp is _____.

    (a) 1.35   (c) 0.77

    (b) 0.92   (d) 0.51

ANSWER SHEET

SHEET METAL AND VENTILATION

1 - B	31 - B	61 - C
2 - C	32 - B	62 - B
3 - B	33 - B	63 - D
4 - A	34 - A	64 - C
5 - C	35 - B	65 - C
6 - A	36 - A	66 - B
7 - B	37 - B	67 - C
8 - A	38 - A	68 - B
9 - A	39 - B	69 - A
10 - D	40 - A	70 - A
11 - C	41 - A	71 - C
12 - A	42 - C	72 - C
13 - B	43 - D	73 - A
14 - A	44 - B	74 - B
15 - C	45 - C	75 - C
16 - B	46 - A	76 - A
17 - A	47 - D	77 - B
18 - B	48 - A	78 - A
19 - C	49 - C	79 - A
20 - A	50 - A	80 - B
21 - A	51 - D	81 - D
22 - A	52 - B	82 - A
23 - A	53 - A	83 - B
24 - B	54 - C	84 - A
25 - A	55 - D	85 - C
26 - A	56 - C	86 - C
27 - B	57 - D	87 - C
28 - A	58 - D	88 - B
29 - C	59 - B	89 - D
30 - A	60 - C	

CLOSED BOOK
4 HOUR EXAMINATION

1. Find the length of "A" for the layout shown in Figure 2.4
    using 45° fittings.

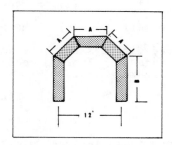

FIGURE 2.4

(a)   59.6 in.   (c)   32.5 in.

(b)   27.6 in.   (d)   31 in.

In Figure 2.5, find the angle of cut for a 90° turn.

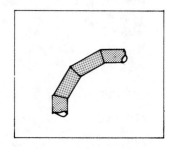

FIGURE 2.5

(a)   45°   (c)   22.50°

(b)   30°   (d)   15°

3. A pipe reducer where the center line of the larger
    pipe is out line with the center of the smaller pipe
    is known as:

    (a) Concentric reducer   (c) Eccentric reducer

    (b) Out of line reducer   (d) Bell reducer

4. A pipe fitting shaped like an ell, but with one female
    end and one male end is called a:

    (a) Male union ell   (c) Street ell

    (b) Female union ell   (d) Female to male ell

5. Threaded and tapped fittings that are screwed into the
    end of other fittings or valves to reduce the size of
    the end openings are known as:

    (a) Reducers   (c) Bushings

    (b) Nipples   (d) Increasers

6. Which is the proper abbreviations for the fitting as
    shown in Figure 2.6, the fitting being Standard Black
    Cast Iron Screwed 125# working pressure:

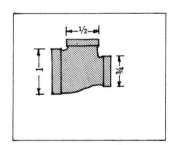

FIGURE 2.6

    (a) 1/2" x 1" x 3/4" std. blk ci scd tee 125# W.P.

    (b) 1" x 1/2" x 3/4" std. blk ci scd tee 125# W.P.

    (c) 3/4" x 1" x 1/2" std. blk ci scd tee 125# W.P.

    (d) 1" x 3/4" x 1/2" std. blk ci scd tee 125# W.P.

7. When making up a pipe joint where you are putting a valve on an 8" long pipe nipple, it is considered good practice to:

    (a) Put the valve in the vise and make up nipple with pipe wrench

    (b) Put the pipe nipple in the vise and make up valve with pipe wrench

    (c) You may either put valve or nipple in vise and make up other with pipe wrench

    (d) Don't use vise to make up this kind of joint, use 2 pipe wrenches

8. What kind of threads does a street ell have:

    (a) External threads only

    (b) Internal threads only

    (c) Both external and internal threads

    (d) NF and NP threads

9. On welded pipe fittings the center to face dimension of a standard 45° 6" weld ell is:

    (a) 3-3/4"   (c) 3"

    (b) 5"   (d) 6"

10. What fittings should be used to connect 1/4" O.D. copper tubing to a 1/4" internal pipe thread opening in a machine body:

    (a) A union   (c) A tee fitting

    (b) A half-union   (d) A street ell

11. Find the length of a piece of pipe for a 90° bend with a radius of 40 in. and with two 15 in. tangents.

    (a) 65.82 in.   (c) 77.82 in.

    (b) 92.82 in.   (d) 62.82 in.

12. Which valve is most suitable for throttling flow:

    (a) Globe   (c) Check

    (b) Gate   (d) Swing

13. Flow through swing check valve is in a straight line and without restriction at the seat. This effect on flow is the reason for generally using checks in combination with _____ valves.

    (a) Globe   (c) All

    (b) Lift check   (d) Gate

14. Pop safety valve for steam air or gas should always be installed with the stem in a horizontal position.

    (a) True   (b) False

15. On the installation of a pressure regulator for use in process work you should always install a relief valve and a pressure gauge on the high pressure side of the regulator.

    (a) True   (b) False

16. For the ordinary installation what is considered good rule of thumb practice in spacing at intervals pipe hangers or supports for pipe:

    (a) 5 feet   (c) 10 feet

    (b) 20 feet   (d) 30 feet

17. In designating the outlets of reduced fittings, whether of the flanged or screwed type, the openings should be read in a certain order. Therefore, on side outlet reducing fittings the size of the side outlet is named in what order:

    (a) First   (c) First or last

    (b) According to size   (d) Last

18. A wrap around as used in pipe fitting is used for what purpose:

   (a) As a protective device

   (b) To draw a straight line around pipe

   (c) For insulation

   (d) To rig pipe for lifting

19. In pipe bending a general rule for finding the length of any bend is as follows:

   (a) R^2 x Angle in Degrees x 0.0175

   (b) Radius x Angle in Degrees x 0.0175

   (c) R^2 x Angle in Degrees x 3.1416

   (d) Radius x Angle in Degrees x 3.1416

20. Which pipe has the thicker wall:

   (a) Schedule 20

   (b) Schedule 40

   (c) Schedule 80

   (d) Schedule 30

21. What is the formula for finding the area of a circle:

   (a) $A = \pi D^2$ (c) $A = \pi R$

   (b) $A = \pi R^2$ (d) $A = \pi D$

22. The plastic flow of pipe within a system is known as:

   (a) Corrosion   (c) Stress

   (b) Strain   (d) Creep

23. A method of joining metals using fusible alloys having melting points under 700° F:

    (a) Acetelyne welding   (c) Soldering

    (b) Brazing   (d) Arc welding

24. To find the volume of the space within the walls of the pipe:

    (a) $V = \pi R^2$ (c) $V = \text{Area} \times \pi R^2$

    (b) $V = \text{Length} \times \pi R^2$ (d) $V = \text{Diameter} \times \pi R^2$

25. All pipe fittings have 3" diameter for pressures above 250 psi, but not above 400 psi shall:

    (a) Have screw or flanged ends

    (b) Have screw or welding ends

    (c) Have flanged or welding ends

    (d) Have screw, flanged or welding ends

26. On piping systems all threads shall conform to:

    (a) American Standard for taper pipe threads

    (b) National Standards for taper pipe threads

    (c) International Standard for taper pipe threads

    (d) ASME Standards

27. Where steam pipes pass through walls, partitions, etc., constructed of combustible material, protecting metal sleeves or thimbles shall be provided to give a clearance all around the pipe and covering of not less than:

    (a) 1/2"   (c) 1/8"

    (b) Snug fit   (d) 1/4"

28. The seam on any ferrous pipe which is being bent should be located at:

    (a) Top   (c) Far side

    (b) Bottom   (d) Near side

29. What material may not be used for hanger rods, turn-buckles or clamps:

    (a) Malleable iron   (c) Cast iron

    (b) Brass   (d) Stainless steel

30. Welding qualification tests on ferrous pipe and materials shall be in accordance with the provisions of:

    (a) American Pressure Vessel Code

    (b) ASME Boiler and Pressure Vessel Code

    (c) Pittsburgh Testing Laboratory

    (d) Hartford Boiler

31. On pressure piping, branch connections made by welding couplings or half couplings directly to the main pipe shall not be used for branches larger than how many inches in nominal pipe size:

    (a) 1"   (c) 3"

    (b) 2"   (d) 4"

32. On pressure piping screw threads shall conform to:

    (a) 8 threads per inch

    (b) Size of fittings

    (c) U.S. Standard for taper pipe threads

    (d) American Standard for taper pipe threads

33. A straight line drawn from the center to the extreme edge of a circle is known as:

    (a) Circumference of a circle

    (b) Area of a circle

    (c) Diameter of a circle

    (d) Radius of a circle

34. Pipe with its axis in the horizontal plane and rolled during welding so that the weld metal is deposited from the top and within plus or minus, 15 degrees from the vertical place is known as what position weld:

    (a) Pipe-horizontal rooled; Weld-flat position

    (b) Pipe-horizontal fixed; Weld-flat overhead position

    (c) Pipe-vertical fixed; Weld-horizontal position

    (d) Pipe-fixed weld; Vertical or horizontal position

35. The difference between the pressure in a piping system and atmospheric pressure is measured in:

    (a) Pounds per square inch   (c) Total pressure

    (b) Absolute pressure   (d) Inches of vacuum

36. On industrial gas and air piping, what should be installed on each side of and immediately adjacent to every pressure reducing device:

    (a) A sight glass   (c) A valve

    (b) A gauge   (d) An orifice plate

37. On industrial gas and air piping, underground cast-iron pipe shall be installed (unless prevented by other underground structures) with a minimum cover of:

    (a) 24"   (c) 18"

    (b) 36"   (d) 12"

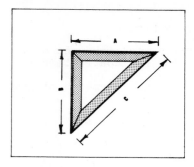

FIGURE 2.7

38. Figure 2.7 shows a 45°angle iron bracket in which the "B" dimension is 24 inches. What is the "C" dimension?

   (a)   36 in.   (c)   38 in.

   (b)   32-1/2 in.   (d)   34 in.

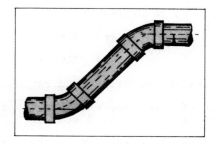

FIGURE 2.8

39. Figure 2.8 shows a 45° offset in which the *travel* is 8-1/2 inches, therefore the *run* is _____.

   (a)   10 in.   (c)   7 in.

   (b)   8-1/2 in.   (d)   6 in.

40. What is the purpose of an expansion joint:

    (a) To minimize vibration

    (b) To allow for normal movement of the pipe due to expansion and contraction

    (c) To compensate for movements of pumps or compressors

    (d) Make removal of pipe easy

41. The treatment of liquids under pressure and at rest is:

    (a) Hydrostatics   (c) Hydrophonics

    (b) Hydraulics   (d) Compression

42. The treatment of liquids under pressure and in motion is:

    (a) Hydrostatics   (c) Hydrophonics

    (b) Hydraulics   (d) Compression

43. Paper vegetable fiber or rubber gaskets shall not be used for temperatures in excess of:

    (a) 212° F   (c) 316° F

    (b) 250° F   (d) 280° F

44. What factors determine the size of an expansion tank?

    (a) The amount of space the water in the system requires in its expanded state

    (b) The amount of space the air in the system requires in its expanded state

    (c) The amount of air in system

    (d) The operating pressure in system

45. You should always use a full face gasket on a:

    (a) Ring type joint flange

    (b) Raised face flange

    (d) Flat face flange

46. When the pitch of the thread on a pipe is 1/4" how many turns are required to thread 2-1/2" of the pipe?

    (a) 8   (c) 12

    (b) 10   (d) 13

47. Two pipes with an area of 3 sq in. and 4 sq in. respectively discharge into a single header. What is the diameter of the header if it has an area = to the sum of the area of the two pipes?

    (a) 4"   (c) 3"

    (b) 6"   (d) 5"

48. Resistance to flow in a piping system can be decreased most satisfactory by:

    (a) Increasing pipe size

    (b) Increasing the pressure

    (c) Increasing the velocity

    (d) Decreasing the pipe size

49. What fitting should be used to connect 1/4" O.D. copper tubing flared to a 1/4" internal pipe thread opening in a water pump body?

    (a) A union   (c) A tee fitting

    (b) A half union   (d) A street ell

50. Type K copper tube has a greater wall thickness than type L copper tube.

    (a) True   (b) False

PRESSURE & PROCESS PIPING

ANSWER SHEET

1 - A	26 - A
2 - D	27 - D
3 - C	28 - A
4 - C	29 - C
5 - C	30 - B
6 - D	31 - C
7 - B	32 - D
8 - C	33 - D
9 - A	34 - A
10 - B	35 - A
11 - B	36 - C
12 - A	37 - B
13 - D	38 - D
14 - B	39 - D
15 - B	40 - B
16 - C	41 - A
17 - D	42 - B
18 - B	43 - B
19 - B	44 - A
20 - C	45 - C
21 - B	46 - B
22 - D	47 - C
23 - C	48 - A
24 - B	49 - B
25 - C	50 - A

3. CODE QUESTIONS

CODE QUESTIONS

National and Local
(Heating, air conditioning, refrigerating and piping)

The following questions are drawn from several representative codes to present an overview of the typical code questions encountered in most examinations. Although the codes show a great similarity, there are differences between various local areas in some cases. The candidate for license examination should certainly own a copy of the current *Code Book* that rules in his particular locale, and be familiar with how the information is organized and presented.

Abbreviations for reference numbers are as follows:

ANSI	B31.1.0	Power Piping (Code for Pressure Piping, ASME, 1967)
ANSI	B31.5	Refrigeration Piping (Code for Pressure Piping, ASME, 1966)
ANSI	B31.8	Gas Transmission and Distribution Piping Systems (ASME, 1968)
ANSI	B9.1	Safety Code for Mechanical Refrigeration (ASHRAE, 1971)
ANSI	A13.1	Scheme for the Identification of Piping Systems (ASME, 1956)
ASME	Section IV	Heating Boilers: Boiler & Pressure Vessel Code (ASME, 1968)
NFPA		National Fire Codes (1971; listed by pamphlet numbers)
SSBC		Southern Standard Building Code (1971)
SFBC		South Florida Building Code (1973)
NYCBC		New York City Building Code (1970)
LAC		Los Angeles Code for HVAC (1972)
Note:		The Los Angeles Code is essentially the same as the Uniform Mechanical Code (ICBO)

Answers to code questions may be found at the end of this chapter following the excerpts from national standards.

REFERENCE	QUESTIONS ANSI

ANSI B9.1
2.59

1. A sealed absorbtion system is a unit system for Group _____refrigerants.

(a) 1   (c) 3
(b) 2   (d) any of these groups

ANSI B9.1
8.4.2.2

2. Soft copper tubing may be used for refrigerant piping _____inches or less.

(a) 1-1/8   (c) 1-3/8
(b) 1-1/4   (d) 1-5/8

ANSI B9.1
5.1

3. Refrigerant 22 may be classified as belonging to Group _____.

(a) 1   (c) 3
(b) 2   (d) 4

ANSI B9.1
8.4.4

4. Sweat joints on copper tubing in refrigerating systems must be brazed on systems using Group _____refrigerants.

(a) 1   (c) 2
(b) 1 and 2   (d) 2 and 3

ANSI B9.1
5.1

5. Ammonia may be classified as belonging to Group _____.

(a) 1   (c) 3
(b) 2   (d) 2 and 3

ANSI B9.1
8.1.2

6. Aluminum zinc or magnesium shall not be used in contact with _____in a refrigerating system.

(a) CH_3CL (c) $CHCL_2F$
(b) CCL_2F_2 (d) $CHCL_f2$

ANSI B9.1
5.1

7. Propane may be classified as belonging to _____.

(a) Groups 1 and 2
(b) Groups 2 and 3
(c) Groups 1 and 3
(d) None of these

REFERENCE	QUESTIONS ANSI (Continued)

ANSI B9.1
10.1

8. Pressure vessels of ____ psi or greater, shall be equipped with pressure-relief devices.

    (a) 79.6   (c) 212
    (b) 112   (d) All pressure
                           vessels

ANSI B9.1
10.4.8

9. Discharge of pressure-relief device and possible plugs on systems containing more than 6 lbs of Group 2 or Group 3 refrigerants shall be _____.

    (a) Above 7-1/2 ft high
    (b) Below 10 in. from floor
    (c) To the outside of building
    (d) Into approved safe waste

ANSI B9.1
12.2

10. Which of the following may not be used within a refrigerant system for testing:

    (a) NH_3 (c) R-12
    (b) Oxygen   (d) Sulphur dioxide

ANSI B31.5
Table
521.3.5

11. The minimum stock size of hanger rod used to support galvanized pipe exposed to weather which is 2 inches or less (nominal pipe size) shall be _____ diameter.

    (a) 1/8"   (c) 3/8"
    (b) 1/4"   (d) 1/2"

ANSI B31.5
Table
521.3.5

12. The minimum stock size of hanger rod used to support galvanized pipe exposed to weather that is over 2 inches (nominal pipe size) shall be _____ diameter.

    (a) 1/8"   (c) 3/8"
    (b) 1/4"   (d) 1/2"

ANSI B31.5
Table
523.2.3

13. Cast iron and malleable iron shall not be used for piping components in hydrocarbon service at pressures above ____ psi.

    (a) 210   (c) 380
    (b) 300   (d) 460

REFERENCE	QUESTIONS ANSI (Continued)

ANSI B31.5
Table
537.2.2

14. When testing refrigerant piping systems after erection, test pressure must be maintained for _____ minutes without loss of pressure.

(a) 30   (c) 120
(b) 60   (d) 210

ANSI B31.5
Table
537.2.2

15. The minimum field refrigeration leak test pressure, (high side) for dichlorodifluoromethane is _____ psig.

(a) 300   (c) 253
(b) 285   (d) 325

ANSI B31.5
500.2

16. A _____ is a groove melted into the base metal adjacent to the toe of a weld metal.

(a) Slag inclusion
(b) Weldment
(c) Undercut
(d) Fillet

ANSI B31.5
500.2

17. In piping terminology, which of the following is *not* considered a fixture.

(a) Clevis   (c) Saddle
(b) Strut   (d) Turnbuckle

ANSI B31.8
831.33

18. In a gas distribution system, mechanical fittings may be used for making hot taps on pipelines if they are designed at _____ the operating pressure of the pipe.

(a) Equal to   (c) 10% over
(b) 30% over   (d) 50% over

ANSI B31.8
831.33

19. All pipelines and mains, of gas distribution systems, operating at a hoop strength of 30% or more of the specified minimum yield strength must be tested _____.

(a) After construction and before operation
(b) After construction and after operation
(c) Before construction or operation
(d) For permeability

REFERENCE	QUESTIONS ANSI (Continued)

ANSI B31.8
841.441

20. Any gas distribution pipeline which will operate at less than 100 psi shall be leak tested _____.

   (a) After construction and before operation
   (b) After construction and after operation
   (c) Before construction or operation
   (d) For permeability

REFERENCE	QUESTIONS ASME

ASME
Sect. IV
HG-400.1

21. A steam boiler shall have one or more spring-pop safety valve set to discharge at a pressure of not to exceed ____ psi.

   (a) 5   (c) 15
   (b) 10   (d) 25

ASME
Sect. IV
HG-530.1

22. _____type boilers must show the type H-symbol.

   (a) Water-tube   (c) split section
   (b) Fire-tube   (d) all of these

ASME
Sect. IV
HG-602

23. Water-gage glasses must have a blow down valve with a _____ inch minimum diameter opening.

   (a) 1   (c) 1/2
   (b) 3/4   (d) 1/4

ASME
Sect. IV
E-100

24. A boiler which has a self-supporting water cooled shell or furnace bottom is called a _____ boiler.

   (a) wet back   (c) water cooled
   (b) wet bottom   (d) water tube

ASME
Sect. IV
E-100

25. The difference between the opening and closing pressure of a safety or relief valve is know as the _____.

   (a) Differential pressure
   (b) Pressure drop
   (c) Blow down
   (d) P-orifice

REFERENCE	QUESTIONS SFBC

SFBC
4102.4 (m)

26. Underground tanks for storing flammable liquids must have a vent pipe draining to the tank; the top of such vent pipe shall not be closer than _____ ft to any building.

    (a) 3   (c) 7
    (b) 5   (d) 10

SFBC
4903.7

27. Where an air conditioner is serving a building of Group 1 occupancy, the outside air intake must be a minimum of _____ ft from any plumbing system vent terminal.

    (a) 3   (c) 7
    (b) 5   (d) 10

SFBC
4103.3 (b)

28. Hoods over restaurant equipment shall not be raised more than _____ ft from the floor.

    (a) 6-1/2   (c) 7-1/2
    (b) 7   (d) 8

SFBC
4103.3 (d)

29. Duct systems conveying exhaust from restaurant hoods shall operate at a velocity no less than _____ fpm.

    (a) 1500   (c) 2000
    (b) 1700   (d) 2200

SFBC
4103.3 (d)

30. A restaurant kitchen exhaust hood is 6 ft long and 4 ft wide, what is the minimum air volume required in cfm?

    (a) 100   (c) 1400
    (b) 1000   (d) 2400

SFBC
3703.6 (b)

31. In a fire protected ceiling, duct openings must be protected by a fire damper if the opening is larger than _____ in each 100 sq ft.

    (a) Damper not required
    (b) 100 sq in.
    (c) 150 sq in.
    (d) 200 sq in.

REFERENCE	QUESTIONS SFBC (Continued)

SFBC
3801.1 (c)

32. An approved fire sprinkler system is required in buildings of Group E occupancy having an area of more than _____ sq ft.

    (a) 5,000   (c) 1,500
    (b) 10,000   (d) 15,000

SFBC
4606.7

33. Condensate drains for air conditioning systems shall be a minimum of _____ inch diameter if the system is not over 10 tons.

    (a) 1/2   (c) 1
    (b) 3/4   (d) 1-1/4

SFBC
4616.4 (d)

34. A supply well for an air conditioning condenser system must be located at least _____ horizontally from a septic tank or drain field.

    (a) 25 yds   (c) 25 ft
    (b) 10 ft   (d) 40 ft

SFBC
4802 (d)

35. A toilet room without air conditioning is 10 ft long x 8 ft wide x 10 ft high, what is the minimum ventilation rate required in cfm?

    (a) 160   (c) 642
    (b) 267   (d) 800

SFBC
4802 (c)

36. A dry cleaning plant is 40 ft long x 30 ft wide x 10 ft high, what is the minimum ventilation rate required in cfm?

    (a) 12,000   (c) 2,000
    (b) 6,000   (d) 4,000

SFBC
4006.6

37. A Certificate of Inspection for a high pressure boiler shall be valid for a period of not more than _____.

    (a) 6 months   (c) 24 months
    (b) 12 months (d) 36 months

SFBC
4903.7

38. All air conditioning systems serving a commercial establishment shall be provided with a minimum of _____ cfm of outside air for each person in the conditioned space.

REFERENCE	QUESTIONS SFBC (Continued)

(a) 1   (c) 7-1/2
(b) 5   (d) 10

SFBC
4903.4

39. All refrigerating system shall be maintained in a clean condition by _____.

(a) A licensed service company
(b) The user
(c) The owner
(d) The installing contractor

SFBC
3902.1

40. Solid and liquid fuel burning equipment require Class _____ venting.

(a) A   (c) C
(b) B   (d) None of these

SFBC
3902.3 (c)

41. Metal smoke stacks 12 in. to 16 in. diameter shall be not less _____ gauge.

(a) 26   (c) 12
(b) 22   (d) 10

REFERENCE	QUESTIONS SSBC

SSBC
2001.6 (b)

42. A toilet room without air conditioning is 10 ft long x 8 ft wide x 10 ft high, what is the required minimum ventilation rate in cfm?

(a) 160   (c) 642
(b) 267   (d) 800

SSBC
501.2 (j)

43. A dry cleaning plant is 40 ft long x 30 ft wide x 10 ft high, what is the minimum ventilation rate required in cfm?

(a) 12,000   (c) 2,000
(b) 6,000   (d) 4,000

REFERENCE	QUESTIONS SSBC (Continued)

SSBC
824

44. A high velocity round duct 23 in. diameter with a longitudinal seam construction, must be at least _____ gauge sheet steel.

(a) 18   (c) 22
(b) 20   (d) 24

SSBC
819.4

45. A steam or hot water boiler must have a _____.

(a) Drain valve
(b) Individually controlled make up valve
(c) None of these
(d) Both of these

SSBC
819.6

46. An automatically fired steam boiler must have a _____.

(a) High limit shutoff
(b) Separate operating shutoff
(c) None of these
(d) Both of these

SSBC
811-A

47. Low pressure, oil fired boilers must have a minimum front clearance of _____ inches.

(a) 16   (c) 24
(b) 20   (d) 48

SSBC
804.3 (c)

48. The minimum clearance for any metal chimney is ____ inches from a wall of a concrete block constructed building.

(a) 2   (c) 4
(b) 3   (d) 6

SSBC
814.1

49. The requirements for restaurant kitchen exhaust shall be governed by the _____.

(a) SFBC - 1971
(b) NFPA -96.1971
(c) National Building Code -1972
(d) Federal Pollution Commission

REFERENCE	QUESTIONS NFPA

NFPA - 32
2301

50. A dry cleaning plant is 40 ft long x 30 ft wide x 10 ft high, what is the minimum ventilation rate required in cfm?

    (a) 12,000   (c) 2,000
    (b) 6,000   (d) 4,000

NFPA - 96
Appendix A(E)

51. Commercial cooking equipment grease filters must be figures at no less than _____ inches of static when calculating exhaust fan friction loses.

    (a) 0.08   (c) 0.1
    (b) 0.06   (d) 0.2

NFPA -90A
906

52. The location of all automatic fire dampers shall be marked on all plans by the _____.

    (a) Designer of an air duct system
    (b) Building & zoning department
    (c) Installing contractor
    (d) Local fire authority

NFPA -90A
403

53. Fresh air intakes must have the following:

    (a) 1/2 in. mesh screen
    (b) Spark proof construction
    (c) Fusible link
    (d) Non-combustible grille

NFPA - 30
11

54. Any liquid with a flash point below _____ and vapor pressure not over 40 psi shall be considered a flammable liquid.

    (a) 120 F   (c) 160 F
    (b) 140 F   (d) 212 F

NFPA - 31
1001

55. A high pressure boiler furnishes steam at more than _____ psig.

    (a) 30   (c) 20
    (b) 25   (d) 15

REFERENCE	QUESTIONS NFPA (Continued)

NFPA - 31
1405

56. The free area of a full wood louvered door measuring 3'-0" x 6'-6" is approximately _____ ft.

    (a) 19.5   (c) 9.0
    (b) 14.5   (d) 5

NFPA - 31
2024

57. Above ground steel tanks for flammable liquids may not exceed _____ gallons capacity.

    (a) 100   (c) 2500
    (b) 250   (d) 3000

NFPA - 90A
303 (B)

58. Flexible connections between air handlers and ducts shall have a melting point of no less than _____ .

    (a) 1200 F   (c) 1700 F
    (b) 1500 F   (d) 2000 F

NFPA - 90A
1002

59. Fire stat fan shutoff must be provided on all air conditioning systems designed for _____ cfm.

    (a) 2,000 to 15,000
    (b) 5,000 to 20,000
    (c) Over 20,000
    (d) None of these

NFPA - 90A
1003

60. Smoke detector fan shutoff must be provided on all air conditioning systems designed for _____ cfm.

    (a) 5,00 to 20,000
    (b) Over 15,000
    (c) 10,000 to 20,000
    (d) Over 20,000

NFPA - 90B
121

61. The minimum gauge galvanized sheet for residential system ductwork is _____ .

    (a) 24   (c) 28
    (b) 26   (d) Neither of these

REFERENCE	QUESTIONS NFPA (Continued)

NFPA - 90B
181 (2)

62. Return ducts in residential warm air systems may be constructed of _____ thick wood.

    (a)   Wood cannot be used
    (b)   1/2 in.
    (c)   3/4 in.
    (d)   1 in.

NFPA - 97M

63. A low-water cutoff is a de-energizer set to trip when the _____.

    (a)   Low-water reaches 180 F
    (b)   Boiler water drops
    (c)   Low-water rises
    (d)   Blow down opens

REFERENCE	QUESTIONS NYCBC

NYCBC
C26-1206-3

64. The actual outdoor air supply for conditioned spaces shall not be less than _____ cfm.

    (a) 3   (c) 7
    (b) 5   (d) 10

NYCBC
C26-1207.3

65. Mechanical exhaust systems for toilets consisting of one water closet or urinal shall exhaust at least ____ cfm.

    (a) 20   (c) 40
    (b) 30   (d) 50

NYCBC
C26-1208.3

66. Ductwork serving dwelling units shall be lined with sound attenuating material for at least _____ on either side of the fan.

    (a) 7 diameters (c) 15 ft
    (b) 10 diameters (d) 20 ft

NYCBC
C26-1208.3

67. The minimum thickness of vibration isolators for fan equipment mounted on roofs shall be _____ in.

    (a) 1/4   (c) 3/4
    (b) 1/2   (d) 7/8

REFERENCE	QUESTIONS NYCBC (Continued)

NYCBC
C26-1301.2

68. Any ventilation system that exhausts injurious substances or may create a fire hazard is subject to controlled inspection by _____.

   (a) City inspector
   (b) Fire department
   (c) Architect or engineer
   (d) Test and balance agent

NYCBC
C26-1416.1

69. On fuel oil storage tanks where a shut-off valve is provided in the supply line, such valve must be rated at a minimum of _____ psi.

   (a) 100   (c) 135
   (b) 125   (d) 150

NYCBC
C26-RS-13-1
315

70. Any duct passing through two or more floors or a floor and roof and having a cross-sectional area of more than 2 sq ft, shall have a ____ hr fire rating.

   (a) 1/2   (c) 1-1/2
   (b) 1   (d) 2

NYCBC
C26-RS-13-1
319

71. For ducts with a cross-sectional area of over 2 sq ft, the hangers shall be constructed of minimum _____ strap.

   (a) 1 in. x 1/8
   (b) 1-1/4 in. x 1/8
   (c) 1-1/4 in. x 1/16
   (d) 1 in. x 1/32

NYCBC
C26-RS-13-1
403

72. Fresh air intakes shall be protected by screens of corrosion-resistant material not larger than _____ mesh

   (a) 1/4 in.   (c) 1/2 in.
   (b) 3/8 in.   (d) 5/8 in.

REFERENCE	QUESTIONS NYCBC (Continued)

NYCBC
C26-RS-13-3

73. The standard for restaurant cooking equipment is _____.

(a) AGA-319   (c) NESCA-96
(b) NFPA-90B   (d) NFPA-96

REFERENCE	LOS ANGELES CODE FOR HVAC (1972 ed.)

LAC 95.1230

74. A permit for refrigeration system installation is valid for a period of _____.

(a) 30 days   (c) 6 mos.
(b) 90 days   (d) 1 year

LAC 95.1261

75. Contractor plans must show the weight of any equipment in excess of ___ lbs.

(a) 150   (c) 200
(b) 175   (d) 250

LAC 95.3111

76. Oil burning appliances may be connected to the fuel source with a listed flexible metal hose providing such connection is _____.

(a) Above ground
(b) 1/4" pitched
(c) Over 1/2" diameter
(d) Under 3' long

LAC 95.3140

77. All warm air furnaces must have a temperature limit control on the downstream air side with a maximum setting of _____ degrees.

(a) 200   (c) 210
(b) 250   (d) 220

LAC 95.3210

78. Fusible link fire dampers shall _____combustion air opening or duct.

(a) Not be located in any
(b) Be located in every
(c) Have a melt point of 165°F in the
(d) Have a melt point of 286°F in the

REFERENCE	LOS ANGELES CODE FOR HVAC (Continued)

LAC 95.3350   79. The minimum circulating air supply
opening to a 50 MBtu/hr warm air
furnace is _____ .

    (a) 75 sq in.   (c) 144 sq in.
    (b) 100 sq in.   (d) 150 sq in.

LAC 95.5180   80. A restaurant kitchen exhaust hood is
6 ft long and 4 ft wide, what is the
minimum air volume required in cfm?

    (a) 1200   (c) 2400
    (b) 1800   (d) 3600

LAC 95.5180   81. Hoods over restaurant cooking equipment
calculated from the formula Q= 50 PD,
shall have a maximum "D" of ____ inches.

    (a) 36   (c) 42
    (b) 48   (d) 54

LAC 95.6172   82. Comfort cooling systems, other than
dwellings, shall have no less than
_____ cfm outside air per occupant of
conditioned space.

    (a) 1   (c) 7-1/2
    (b) 10   (d) 5

LAC 95.6172   83. Residential comfort cooling systems
shall not exceed ____ fpm velocity
at return air grilles.

    (a) 450   (c) 400
    (b) 600   (d) 500

LAC 95.7150   84. The minimum size condensate drain pipe
for any absorption system shall be
____ in.

    (a) 1/2   (c) 1
    (b) 3/4   (d) 1-1/2

LAC 95.12210   85. NH_3 is classified as a class _____
refrigerant.

    (a) 1   (c) 3
    (b) 2   (d) 4

REFERENCE	LOS ANGELES CODE FOR HVAC (Continued)

LAC 95.12410   86. Every machinery room shall have an exhaust system providing a complete change of air at least _____ times per hour.

      (a) 20   (c) 15
      (b) 17   (d) 12

SPEED-O-GRAPH

AIR VOLUME REQUIRED FOR CANOPY-TYPE
KITCHEN EXHAUST HOODS

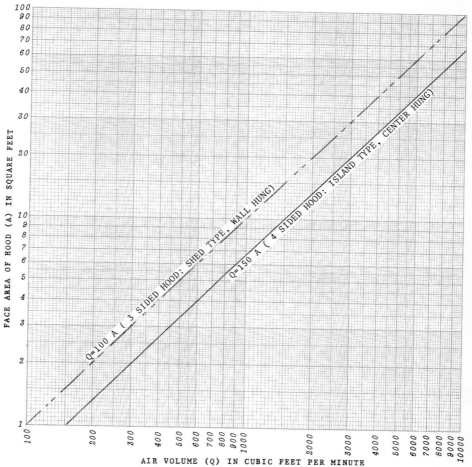

AIR VOLUME (Q) IN CUBIC FEET PER MINUTE
(Based on NFPA Requirements Where Hood Face is Maximum 7 ft From Floor)

FIGURE 3.1

REFERENCE EXTRACTS FROM NATIONAL STANDARDS

The following excerpts from National Standards are extracted by permission of the publishers.

1. *Scheme for the Identification of Piping Systems* (ANSI A13.1 - 1956, with permission of the publisher, The American Society of Mechanical Engineers, 345 East 47 St., New York, N.Y.

2. *Refrigeration Piping Code* (ANSI B31.5 - 1966) a section of the ANSI Standard Code for Pressure Piping, with permission of the publisher, The American Society of Mechanical Engineers.

3. *Safety Code for Mechanical Refrigeration* (ANSI B9.1 - 1971) with permission of the publishers, The American Society of Heating, Refrigerating & Air Conditioning Engineers, 345 East 47 St., New York, N.Y., and The Air Conditioning and Refrigeration Institute.

NOTE: ANSI Standards are subject to revision every five years. Persons referring to these Standards should check dates on reference material for current issues.

American Standard

Scheme for the
Identification of Piping Systems

Object and Scope

1 The scheme is intended to establish a common code to assist in the identification of materials conveyed in piping systems and is intended to form an acceptable basis for a universal scheme. The use of this standard will promote greater safety and will lessen the chances of error, confusion or inaction.

2 This scheme concerns only the identification of piping systems in industrial and power plants. It does not cover pipes buried in the ground or electrical conduits.

Definitions

3 Piping Systems. For the purpose of this scheme, piping systems shall include in addition to pipes of any kind: fittings, valves and pipe coverings. Supports, brackets, or other accessories are specifically excluded from applications of this standard. Pipes are defined as conduits for the transport of gases, liquids, semi-liquids or plastics, but not solids carried in air or gas.

4 Fire Protection, Materials and Equipment. This classification includes sprinkler systems and other fire-fighting or fire protection equipment. The identification for this group of materials may also be used to identify or locate such equipment as alarm boxes, extinguishers, fire blankets, fire doors, hose connections, hydrants and any other fire-fighting equipment.

5 Dangerous Materials. This group includes materials which are hazardous to life or property because they are easily ignited, toxic, corrosive at high temperatures and pressures, productive of poisonous gases or are in themselves poisonous. It also includes materials that are known ordinarily as fire producers or explosives.

6 Safe Materials. This group includes those materials involving little or no hazard to life or property in their handling. This classification includes materials at low pressures and temperatures, which are neither toxic nor poisonous and will not produce fires or explosions. People working on piping systems, carrying these materials run little risks even though the system had not been emptied.

7 Protective Materials. This group includes materials which are piped through plants for the express purpose of being available to prevent or minimize the hazard of the dangerous materials above mentioned. It would include certain special gases which are antidotes, to counteract poisonous fumes, piped for the express purpose of release in case of danger. This classification also covers protective materials for purposes other than for fire protection which is covered under Section 4.

Method of Identification

8 Positive identification of a piping system content shall be by lettered legend giving the name of the content in full or abbreviated form. Arrows may be used to indicate the direction of flow. Where it is desirable or necessary to give supplementary information such as hazard or use of the piping system content, this may be done by additional legend or by color applied to the entire piping system or as colored bands. Legends may be placed on colored bands.

Examples of legend to give both positive identification and supplementary information as regards hazards or use are:

Water	— Fire protection
Ammonia	— Anhydrous — Dangerous liquid and gas
Acetone	— Extremely flammable liquid
Hydrogen	— Extremely flammable gas
Air	— High-pressure gas
Carbon Dioxide	— Fire protection

NOTE: Manual L-1, third revision, 1953, published by Manufacturing Chemists Association, Inc., is a valuable guide in respect to supplementary legend.

9 When color, applied to the entire piping system or as colored bands, is used to give supplementary information it shall conform to the following:

CLASSIFICATION	PREDOMINANT COLOR
F — Fire-protection equipment	Red
D — Dangerous materials	Yellow(or orange)
S — Safe materials	Green (or the achromatic colors, white, black, gray or aluminum)
and, when required,	
P — Protective materials	Bright blue

SCHEME FOR IDENTIFICATION OF PIPING SYSTEMS

10 The above colors have been chosen to identify the main classifications because they are readily distinguishable one from another under normal conditions of illumination. Fig. 1 shows the four main classification color bands, color of the legend letters and their suggested placement location, also the recommended width of color band "A" together with the size of the legend letters "B" for various pipe diameters.

11 Color bands, if used, shall be painted or applied on the pipes to designate to which of the four main classifications its contents belongs. The bands should be installed at frequent intervals on straight pipe runs (sufficient to clearly identify) close to all valves, and adjacent to all change-in-directions, or where pipes pass through walls or floors. The color identification may be accomplished by the use of decals or plastic bands which are made to conform with the standards. If desired the entire length of the piping system may be painted the main classification color.

12 Attention has been given to visibility with reference to pipe markings. Where pipe lines are located some distance above the normal line of the operator's vision the lettering should be placed below the horizontal center line of the pipe as shown in Fig. 1.

13 In certain types of plants it may be desirable to label the pipes at junction points or points of distribution only while at other locations the markings may be installed at necessary intervals all along the piping, close to valves and adjacent to change-in-directions. In any case the number and location of identification markers should be based on judgment for each particular system of piping.

14 Regarding the type and size of letters, the use of stencils of standard sizes ranging in height from ½ in. to 3½ in. is recommended. For identification of pipe less than ¾ in. in diameter the use of a tag is recommended. The lettering or the background may be of the standard color. (See Fig. 1).

15 In cases where it is decided to paint the entire piping, the color and sizes of legend letters stencilled on the piping for identification of material conveyed should conform to the specifications shown in Fig. 1.

Key to Classification Color of Bands-
Color of Legend Letters-
Legend Placement-Width of Color Bands and Size of
Letters for Various Diameter Pipes

KEY TO CLASSIFICATION OF PREDOMINANT COLORS FOR BANDS		COLOR OF LETTERS FOR LEGENDS
F – Fire protection	Red	White
D – Dangerous	Yellow	Black
S – Safe	Green	Black
P – Protective	Blue	White

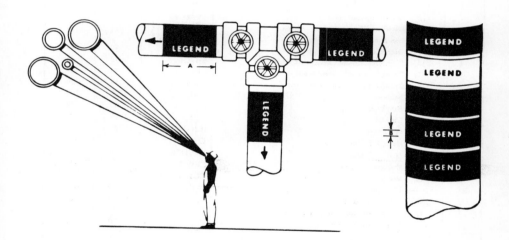

Outside Diameter of Pipe or Covering	Width of Color Band A	Size of Legend Letters B
¾ to 1 ¼	8	½
1 ½ to 2	8	¾
2 ½ to 6	12	1 ¼
8 to 10	24	2 ½
Over 10	32	3 ½

All dimensions are given in inches.

EXCERPTS FROM THE REFRIGERATION PIPING CODE
ANSI B31.5-1966

REFRIGERATION PIPING

Table 537.2.2 Minimum Refrigerant Leak Field Test Pressures

Group	Number	Name	Chemical Formula	Minimum Field Refrigeration leak test pressure, psig	
				High Side	Low Side
II	717	Ammonia	NH_3	300	150
III	600	Butane	C_4H_{10}	95	50
I	744	Carbon Dioxide	CO_2	1500	1000
I	12	Dichlorodifluoromethane	CCl_2F_2	235	140
I	500	Dichlorodifluoromethane 73.8% and Ethylidene Fluoride 26.2%	CCl_2F_2 CH_3-CHF_2 $\}$	285	150
II	1130	Dichloroethylene	$C_2H_2Cl_2$	30	30
I	30	Dichloromethane (Methylene Chloride)	CH_2Cl_2	30	30
I	21	Dichloromonofluoromethane	$CHCl_2F$	70	40
I	114	Dichlorotetrafluoroethane	$C_2Cl_2F_4$	50	50
III	170	Ethane	C_2H_6	1200	700
II	160	Ethyl Chloride	C_2H_5Cl	60	50
III	1150	Ethylene	C_2H_4	1600	1200
III	601	Isobutane	$(CH_3)_3CH$	130	70
II	40	Methyl Chloride	CH_3Cl	210	120
II	611	Methyl Formate	$HCOOCH_3$	50	50
I	22	Monochlorodifluoromethane	$CHClF_2$	300	150
I	13	Monochlorotrifluoromethane	$CClF_3$	685	685
III	290	Propane	C_3H_8	300	150
II	764	Sulphur Dioxide	SO_2	170	85
I	11	Trichloromonofluoromethane	CCl_3F	20	20
I	113	Trichlorotrifluoroethane	$C_2Cl_3F_3$	20	20

Notes:
(a) For refrigerants not listed in Table 537.2.2, the test pressure for the high pressure side shall not be less than the saturated vapor pressure of the refrigerant at 150 F. The test pressure for the low pressure side shall not be less than the saturated vapor pressure of the refrigerant at 110 F. However, the test pressure for either the high or low side need not exceed 125 per cent of the critical pressure of the refrigerant. In no case shall the test pressure be less than 20 psig.

(b) When a compressor is used as a booster to obtain a low pressure and discharges into the suction line of another system, the booster compressor is considered a part of the low side, and values listed under the low side column in Table 537.2.2 shall be used for both high and low side of the booster compressor provided that a low pressure stage compressor of the positive displacement type shall have a pressure relief valve.

(c) In field testing systems using nonpositive displacement compressors, the entire system shall be considered for field test purposes as the low side pressure.

ANSI B 31.5 (continued)

537 TESTS

537.1 Tests before erection or assembly.

537.1.1 Equipment including valves, gauges and regulators, and pipe, tube and fittings are tested by the manufacturer according to standards or applicable specifications, and no further test of equipment or material before installation shall be required.

537.2 Tests after erection.

537.2.1 All refrigerant and brine piping erected on the premises, including valves and fittings, shall be tested and proved tight after complete assembly and before insulating and operating.

537.2.2 Tests for refrigerant piping.

(a) Tests shall be at pressure not less than the minimum shown in Table 537.2.2, except as permitted in Paragraph 537.2.2(b).

(b) The testing fluid for systems erected on the premises using Group I refrigerants in which no pipe exceed 3/4 in. OD may be the refrigerant charged into the system at the saturated vapor pressure of the refrigerant at 70 F or higher.

(c) No oxygen or any combustible gas or combustible mixture of gases shall be used for testing within the system. Water should not be used for testing refrigerant piping systems, but if used, it must be completely removed.

(d) The means used to obtain the test pressure shall have either a pressure-limiting device or a pressure-reducing device and a gage on the outlet side.

(e) Test pressure shall be maintained for 30 minutes without loss of pressure.

(f) Final test, evacuation and charging: Applicable only to Groups I and II refrigerants except Carbon Dioxide and Ammonia.

(1) After a satisfactory pressure test, connect temporarily to the system a high-vacuum pump capable of reducing the absolute pressure in the system to a point where any water present will vaporize at a temperature appreciably below ambient temperature and will be withdrawn from the system.

(2) The system shall be evacuated to an absolute maximum pressure of 0.200 inch of mercury. During evacuation the system ambient temperature shall be higher than 35 F.

(3) After this condition is attained, the vacuum pump shall be valved off from the system for a period of at least 12 hours.

(4) The system shall be considered tight, dry and free of air if, at the expiration of this period, the absolute pressure has not increased by more than 0.02 inch of mercury.

(5) If the pressure rise exceeds 0.02 inch of mercury, indicating a leak or the presence of moisture, the pressure test shall be repeated. If no leak is found, evacuation shall be resumed and continued until dryness as well as tightness is established.

(6) When a satisfactory vacuum has been obtained, it shall be broken by the introduction of the refrigerant vapor (not liquid).

Note: In making the vacuum test, the condition of the oil in the vacuum pump is critical and it may have to be replaced during the evacuation for the pump to develop its inherent capabilities.

537.2.3 *Tests for Brine Piping.*

(a) Piping systems for brine shall be tested at one-and-one-half times the design pressure.

(b) A refrigerant used as a brine shall be treated as a refrigerant.

(c) The procedure in Paragraphs 537.2.2(c) and 537.2.2(f) for refrigerant piping should be used for brine piping when the presence of water may be detrimental.

537.2.4 *Limited Charge Systems.*

Limited charge systems shall be tested in accordance with Paragraph 537.2.2 except that limited charge systems equipped with pressure relief devices may be tested and proved tight at a pressure not less than one-and-one-half times the setting of the pressure relief device.

537.2.5 *Pressure Gages.*

Pressure gages shall be checked for accuracy prior to test, either by comparison with master gages or by setting the pointer as determined by a dead-weight pressure gage tester.

537.3 Repair of Joints

(a) Solder joints which leak shall be disassembled, cleaned, reassembled, refluxed and resoldered. Solder joints shall not be reapired by brazing.

(b) Brazed joints which leak may be repaired by cleaning the exposed area, refluxing and rebrazing.

(c) Welded joints which leak should be chipped and rewelded.

(d) Threaded joints which leak shall be unscrewed and remade.

(e) Flanged joints which leak shall be reassembled with a new gasket.

537.3.1 After joints have been repaired, the system shall be retested by the procedure in Paragraph 537.2.2.

ANSI B 31.5 (continued)

Table 521.3.5 Minimum Sizes of Straps, Rods, and Chains for Hangers

Nominal Pipe Size (inches)	Component (Steel)	Minimum Stock Size (inches)	
		Exposed to Weather	Protected from Weather
1 & Smaller	Strap	1/8 Thick	1/16 Thick x 3/4 Wide
Above 1	Strap	1/4 Thick	1/8 Thick x 1 Wide
2 & Smaller	Rod	3/8 Diameter	3/8 Diameter
Above 2	Rod	1/2 Diameter	1/2 Diameter
2 & Smaller	Chain	3/16 Diameter or Equivalent Area	3/16 Diameter or Equivalent Area
Above 2	Chain	3/8 Diameter or Equivalent Area	3/8 Diameter or Equivalent Area
All Sizes	Bolted Clamps	3/16 Thick; bolts 3/8 Diameter	3/16 Thick; bolts 3/8 Diameter

Note: For nonferrous materials, the minimum stock area shall be increased by the ratio of allowable stresses of steel to the allowable stress of the nonferrous material.

EXCERPTS FROM THE SAFETY CODE FOR MECHANICAL
REFRIGERATION B 9.1-1971

8.4 REFRIGERANT PIPING, VALVES, FITTINGS, AND RELATED PARTS

8.4.1 General

Refrigerating *piping*, valves, fittings, and related parts, except those having a maximum internal or external *design pressure* 15 psig or less, *shall* be *listed* either individually or as part of refrigeration equipment by *an approved nationally recognized testing laboratory* or *shall* comply with the rules of Section 5 of the ANSI Code for Pressure Piping where applicable (see 14.10), and to the following requirements:

8.4.2 Specific Minimum Requirements for *unprotected refrigerant* Pipe and Tubing.

8.4.2.1 *Unprotected* watertube size hard copper tubing used for *refrigerant piping* erected on the *premises, shall* conform to ASTM Specification B88 Types K or L (see 14.13) for dimensions and specifications. Copper tubing with an outside diameter of 1/4 in. *shall* have a minimum nominal wall thickness of not less than 0.030 in.

8.4.2.2 *Unprotected* soft annealed copper tubing used for *refrigerant piping* erected on the *premises, shall not* be used in sizes larger than 1-3/8 in. outside diameter. It *shall* conform to ASTM Specification B280. (See 14.14.) Minimum nominal wall thickness of soft *unprotected* annealed copper tubing *shall* be as follows:

Outside Diameter In.	Wall Thickness In.
0.250	0.030
0.375	0.030
0.500	0.032
0.625	0.035
0.750	0.042
0.875	0.045
1.000*	0.050
1.125	0.050
1.250*	0.055
1.375	0.055

*Not included as standard size in ASTM B280

8.4.3 Metal Enclosures or *pipe ducts* for Soft Copper Tubing.

Rigid or flexible metal enclosures *shall* be provided for soft, annealed copper tubing used for *refrigerant piping* erected on the *premises* and containing Group 2 or 3 *refrigerants,* except that no enclosures *shall* be required for connections between *condensing unit* and the nearest riser box, provided such connections do not exceed 6 ft in length.

8.4.4 Joints on copper tubing used in *refrigerating systems* containing Group 2 or Group 3 *refrigerants, shall* be *brazed joints.* Soldered joints shall not be used in such *refrigerating systems.*

SECTION 11.
INSTALLATION REQUIREMENTS

11.1 Foundations and Supports for *condensing units* or *compressor units shall* be of substantial and non-combustible construction when more than 6 in. high.

11.2 Moving *machinery shall* be guarded in accordance with *approved* safety standards. (See 14.3.)

11.3 Clear Space adequate for inspection and servicing of *condensing units* or *compressor units shall* be provided.

11.4 *Condensing units* or *compressor units* with Enclosures *shall* be readily accessible for servicing and inspection.

11.5 Water Supply and Discharge Connections *shall* be made in accordance with *approved* safety and health standards. (See 14.4.)

ANSI B 9.1 (continued)

11.5.1 Discharge water lines *shall not* be directly connected to the waste or sewer systems. The waste or discharge from such equipment *shall* be through an *approved* air gap and trap.

TABLE 6. LENGTH OF DISCHARGE PIPING FOR PRESSURE-RELIEF DEVICES OF VARIOUS DISCHARGE CAPACITIES

Equiv. length of discharge pipe, ft (L)	Discharge capacity in lb of air per min (C) Standard wall iron pipe sizes, in.							
	½	¾	1	1¼	1½	2	2½	3
RELIEF DEVICE SET AT 25 PSIA (P₁)								
50	0.81	1.6	2.9	5.9	8.7	16.3	25.3	43.8
75	0.67	1.4	2.4	4.9	7.2	13.3	20.9	35.8
100	0.58	1.2	2.1	4.2	6.2	11.5	18.0	30.9
150	0.47	0.95	1.7	3.4	5.0	9.4	14.6	25.3
200	0.41	0.8	1.5	2.9	4.4	8.1	12.6	21.8
300	0.33	0.67	1.2	2.4	3.6	6.6	10.5	17.9
RELIEF DEVICE SET AT 50 PSIA (P₁)								
50	1.6	3.3	5.9	11.9	17.4	32.5	50.6	87.6
75	1.3	2.7	4.9	9.7	14.3	26.5	41.8	71.5
100	1.2	2.3	4.2	8.4	12.3	23.0	36.0	61.7
150	0.94	1.9	3.5	6.9	10.0	18.7	29.2	50.6
200	0.81	1.6	2.9	5.9	8.7	16.3	25.3	43.7
300	0.66	1.3	2.5	4.9	7.1	13.3	21.0	35.7
RELIEF DEVICE SET AT 75 PSIA (P₁)								
50	2.4	4.9	8.9	17.9	26.1	48.7	75.9	131.5
75	2.0	4.1	7.3	14.6	21.4	39.8	62.6	107.0
100	1.7	3.5	6.4	12.6	18.5	34.4	54.0	92.6
150	1.4	2.8	5.2	10.3	15.0	28.0	43.8	75.9
200	1.2	2.5	4.4	8.9	13.1	24.4	37.9	65.6
300	0.9	2.0	3.7	7.3	10.7	19.9	31.5	53.5
RELIEF DEVICE SET AT 100 PSIA (P₁)								
50	3.2	6.6	11.9	23.8	31.8	65.0	101.2	175.2
75	2.7	5.4	9.7	19.4	28.6	53.0	83.6	143.0
100	2.3	4.6	8.5	16.8	24.6	45.9	72.0	123.6
150	1.9	3.8	6.9	13.7	20.0	37.4	58.4	101.2
200	1.6	3.3	5.9	11.9	17.5	32.5	50.6	87.6
300	1.3	2.7	4.9	9.7	14.2	26.5	42.0	71.4
RELIEF DEVICE SET AT 150 PSIA (P₁)								
50	4.9	9.9	17.9	35.7	52.3	97.5	151.8	262.8
75	4.0	8.1	14.6	29.2	42.9	79.5	125.4	214.5
100	3.5	6.9	12.7	25.2	36.9	68.9	108.0	185.4
150	2.8	5.7	10.4	20.6	30.0	56.1	87.6	151.8
200	2.4	4.9	8.9	17.8	26.2	48.7	75.9	131.4
300	1.9	4.0	7.4	14.6	21.1	39.7	63.0	107.1
RELIEF DEVICE SET AT 200 PSIA (P₁)								
50	6.5	13.2	23.8	47.6	69.7	130.0	202.4	350.4
75	5.3	10.8	19.4	38.9	57.2	106.0	167.2	286.0
100	4.6	9.2	16.9	33.6	49.2	91.8	144.0	247.2
150	3.8	7.6	13.8	27.4	40.0	74.8	116.8	202.4
200	3.2	6.5	11.8	23.8	34.9	64.9	101.2	175.2
300	2.6	5.3	9.8	19.4	28.4	52.9	81.0	142.8
RELIEF DEVICE SET AT 250 PSIA (P₁)								
50	8.1	16.5	29.8	59.5	87.1	162.5	253.0	437.0
75	6.7	13.5	24.3	48.6	71.5	132.5	209.0	357.5
100	5.8	11.6	21.2	42.0	61.6	114.8	180.0	309.0
150	4.7	9.5	17.3	34.3	50.0	93.5	146.0	253.0
200	4.4	8.2	14.8	29.7	43.7	81.2	126.5	249.0
300	3.3	6.7	12.3	24.3	35.5	66.2	105.0	178.5
RELIEF DEVICE SET AT 300 PSIA (P₁)								
50	9.7	19.8	35.7	71.4	104.5	195.0	303.6	525.6
75	7.9	16.2	29.1	58.3	85.8	159.0	250.8	429.0
100	6.9	13.9	25.4	50.4	73.9	137.7	216.0	370.8
150	5.6	11.3	20.7	41.1	60.0	112.2	175.2	303.6
200	4.9	9.8	17.8	35.6	52.4	97.4	151.8	262.8
300	3.9	7.9	14.7	29.1	42.6	79.4	126.0	211.2

ANSI B 9.1 (continued)

11.6 Illumination adequate for inspection and servicing of *condensing units* or *compressor units shall* be provided. (See 14.5.)

11.7 Electrical Equipment and Wiring *shall* be installed in accordance with *approved* safety standards. (See 14.6.)

11.8 Gas Fuel Devices and Equipment used with *refrigerating systems shall* be installed in accordance with *approved* safety standards. (See 14.7 and 14.20.)

11.9 Air Duct Systems of air-conditioning equipment for human comfort using mechanical refrigeration *shall* be installed in accordance with *approved* safety standards. (See 14.8 and 14.9.)

11.9.1 *Air ducts* passing through a *Class T machinery room shall* be of tight construction and *shall* have no openings in such rooms.

11.10 Joints and Refrigerant-Containing Parts in *air ducts*. Joints and all refrigerant-containing parts of a *refrigerating system* located in an *air duct* carrying conditioned air to and from a *humanly occupied space shall* be constructed to withstand a temperature of 700 F without leakage into the air stream.

11.11 Exposure of *Refrigerant* Pipe Joints. *Refrigerant* pipe joints erected on the *premises shall* be exposed to view for visual inspection prior to being covered or enclosed.

11.12 Location of *Refrigerant Piping*.

11.12.1 *Refrigerant piping* crossing an open space which affords passageway in any building *shall* be not less than 7-1/2 ft above the floor unless against the ceiling of such space.

11.12.2 Free Passageway *shall not* be obstructed by *refrigerant piping*. *Refrigerant piping shall not* be placed in any elevator, dumbwaiter, or other shaft containing a moving object, or in any shaft which has openings to living quarters or to main *exit hallways. Refrigerant piping shall not* be placed in public *hallways, lobbies*, or stairways, except that such *refrigerant piping* may pass across a public *hallway* if there are no joints in the section in the public *hallway*, and provided nonferrous tubing of 1-1/8 in. outside diameter and smaller be contained in a rigid metal pipe.

11.12.3 *Refrigerant piping shall not* be in-
stalled vertically through floors from one story to another except as follows:
a) It may be installed from the basement to the first floor, or from the top floor to a *machinery* penthouse or to the roof.
b) For the purpose of interconnecting separate pieces of equipment not located on adjacent stories, the *piping* may be carried through intermediate stories in an *approved* rigid and tight continuous fire-resisting *pipe duct* or shaft having no openings into the intermediate stories, or it may be carried on the outer wall of the building provided it is not located in an air shaft, closed court or in other similar spaces enclosed within the outer walls of the building. The *pipe duct* or shaft *shall* be vented to the outside.
c) *Piping* of *direct systems* containing Group 1 *refrigerants* as governed by 6.2.1, need not be enclosed where it passes through space served by that system.

11.12.4 *Refrigerant piping* may be installed horizontally in closed floors or in open joist spaces. *Piping* installed in concrete floors *shall* be encased in *pipe duct*.

11.13 *Machinery Room* Requirements.

11.13.1 Each refrigerating *machinery room shall* be provided with tight-fitting door or doors and have no partitions or openings that will permit the passage of escaping *refrigerant* to other parts of the building.

11.13.2 Each refrigerating *machinery room shall* be provided with means for ventilation to the outer air. The ventilation *shall* consist of windows or doors opening to the outer air, of the size shown in Table 7, or of mechanical means capable of removing the air from the room in accordance with Table 7. The amount of ventilation for *refrigerant* removal purposes *shall* be determined by the *refrigerant* content of the largest system in the *machinery room*.

11.13.3 Air supply and return *ducts* used for *machinery room* ventilation *shall* serve no other area.

11.13.4 Mechanical ventilation, when used, *shall* consist of one or more power-driven exhaust fans, which *shall* be capable of removing from the refrigerating *machinery room* the amount of air specified in Table 7. The inlet to the fan, or fans, or *air duct* connection *shall* be located near the refrigerating equipment. The outlet from the fan, or fans, or *air duct* connec-

ANSI B 9.1 (continued)

tions *shall* terminate outside of the building in an *approved* manner. When *air ducts* are used either on the inlet or discharge side of the fan, or fans, they *shall* have an area not less than specified in Table 7. Provision *shall* be made for the inlet of air to replace that being exhausted.

TABLE 7.
MINIMUM AIR DUCT AREAS AND OPENINGS

Weight of refrigerant in system, lb	Mechanical discharge of air, cfm	Duct area, sq ft	Open areas of windows and doors sq ft
up to 20	150	1/4	4
50	250	1/3	6
100	400	1/2	10
150	550	2/3	12 1/2
200	680	2/3	14
250	800	1	15
300	900	1	17
400	1,100	1 1/4	20
500	1,275	1 1/4	22
600	1,450	1 1/2	24
700	1,630	1 1/2	26
800	1,800	2	28
900	1,950	2	30
1,000	2,050	2	31
1,250	2,250	2 1/4	33
1,500	2,500	2 1/4	37
1,750	2,700	2 1/4	38
2,000	2,900	2 1/4	40
2,500	3,300	2 1/2	43
3,000	3,700	3	48
4,000	4,600	3 3/4	55
5,000	5,500	4 1/2	62
6,000	6,300	5	68
7,000	7,200	5 1/2	74
8,000	8,000	5 3/4	80
9,000	8,700	6 1/4	85
10,000	9,500	6 1/2	90
12,000	10,900	7	100
14,000	12,200	7 1/2	109
16,000	13,300	7 3/4	118
18,000	14,300	8	125
20,000	15,200	8 1/4	130
25,000	17,000	8 3/4	140
30,000	18,200	9	145
35,000	19,400	9 1/4	150
40,000	20,500	9 1/2	155
45,000	21,500	9 3/4	160

11.13.5 *Machinery room, Class T shall* have no flame-producing apparatus permanently installed and operated and also *shall* conform to the following:

a) Any doors, communicating with the building, *shall* be *approved* self-closing, tight-fitting fire doors.

b) Walls, floor, and ceiling *shall* be tight and of not less than one-hour fire-resistive construction.

c) It *shall* have an *exit* door which opens directly to the outer air or through a vestibule-type *exit* equipped with self-closing, tight-fitting doors.

d) Exterior openings, if present, *shall not* be under any fire escape or any open stairway.

e) All pipes piercing the interior walls, ceiling, or floor of such room *shall* be tightly sealed to the walls, ceiling, or floor through which they pass.

f) Emergency remote controls to stop the action of the *refrigerant compressor shall* be provided and located immediately outside the *machinery room.*

g) An independent mechanical ventilation system *shall* be provided and operated continuously.

h) Emergency remote controls for the mechanical means of ventilation *shall* be provided and located outside the *machinery room.*

CODE QUESTIONS

NATIONAL AND LOCAL

ANSWER SHEET

ANSI		SFBC		NFPA		LAC	
1 - B		26 - A		50 - C		74 - D	
2 - C		27 - D		51 - D		75 - C	
3 - A		28 - B		52 - A		76 - D	
4 - D		29 - A		53 - A		77 - B	
5 - B		30 - D		54 - B		78 - A	
6 - A		31 - A		55 - D		79 - A	
7 - D		32 - C		56 - D		80 - C	
8 - D		33 - B		57 - C		81 - B	
9 - B		34 - C		58 - C		82 - D	
10 - B		35 - B		59 - A		83 - D	
11 - C		36 - D		60 - B		84 - B	
12 - D		37 - A		61 - D		85 - B	
13 - B		38 - C		62 - D		86 - D	
14 - A		39 - B		63 - B			
15 - C		40 - A					
16 - C		41 - D		**NYCBC**			

ANSI (continued)

17 - A
18 - A
19 - A
20 - A

SSBC

NYCBC

64 - B
65 - D
66 - D
67 - A
68 - C
69 - B
70 - D
71 - A
72 - C
73 - D

ASME

21 - C
22 - D
23 - D
24 - B
25 - C

SSBC

42 - A
43 - D
44 - B
45 - D
46 - D
47 - C
48 - A
49 - B

4. UNDERSTANDING SPECIAL PROBLEM SOLVING QUESTIONS

COOLING TOWER AND PUMP SELECTION

Recently, questions dealing with cooling tower selection, pump and interconnecting piping have begun to appear in examinations. For some reason such questions did not show up in examination material heretofore. The reason probably has to do with the deletion of this important design step from most manuals and handbooks as well as refrigeration and air conditioning textbooks; and consequently the difficulty in referencing the material.

Although standard pipe data - found in any handbook - may be used, the actual cooling tower and pump selection must be made from a particular manufacturers' catalog or specification sheet. It is well for the license candidate to get several manufacturers' published data for the purpose of making comparisons in ratings and becoming familiar with respective methods. Armed with precise data, selecting the tower and pump, and sizing the interconnecting piping, is quite uncomplicated.

The example below presents an actual problem from a recent examination with only a few minor changes made in the wording. Rather than dealing with the subject in a lengthy essay, the solution is reached in a simple step-by-step manner. As the steps are followed the reader should have no difficulty in understanding the sequence of the procedure and the final solution.

SELECTING THE COOLING TOWER, PUMP, AND INTERCONNECTING PIPING

Given: A balanced system (Figure 4.1) having a total length of 260 ft of straight steel pipe plus 4 gate valves, 11 ells, and 2 tees. Design conditions for the tower are; cold water on 77 F (25 C), hot water off 87 F (30.6 C) and wet bulb 70 F (21.1 C). The system load is 1,200,000 Btu/hr using ref-22.

Find: 1. Gpm required to circulate
2. Economical pipe size
3. Cooling tower size
4. Pump size
5. Bleed-off in gpm
6. Water make-up requirements in gpm

There are several ways to size the cooling tower, pump and pipe; here we present a simple, rapid method to provide the correct answer in the shortest time.

STEP NUMBER 1; *Make a layout of the Tower, Pump, System Condenser, and Interconnecting Pipe, Fittings, and Valves. See Figure 4.1.*

STEP NUMBER 2; *Determine the Amount of Water to be Circulated in Gpm.*

Condensers are usually designed by the manufacturer for a 20 F (11.1 C) rise at 75 F (23.9 C) entering water. At these standard conditions the water requirements are 1.5 gpm/ ton. Where cooling towers are used, the entering water in the summertime is more likely to be about 85 F (29.4 C) This means that the temperature rise (Δt) is more likely to be about 10 F (5.5 C), or one-half of that for street water operation. Therefore twice as much water, or 3 gpm per ton of refrigeration is generally required. Table 4.1 gives approximate condenser water rates for various sytems. If the load as given is 1,200,000 Btu/hr then 1,200,000 ÷ 12,000 Btu/hr = 100 tons of refrigeration. From Table 4.1 select 300 gpm.

TABLE 4.1

APPROXIMATE CONDENSER WATER RATES FOR VARIOUS SYSTEMS

System Description	Water rates in gpm per ton refrigeration
Refrigerant 12 city water	1-1/2 - 2
Refrigerant 22 city water	1-1/2 - 2
Refrigerant 12 cooling tower	3 - 4
Refrigerant 22 cooling tower	3 - 4
Ammonia city water	3
Ammonia cooling tower	4 - 6
Lithium bromide cooling tower	3 -4

SYSTEM   LAYOUT

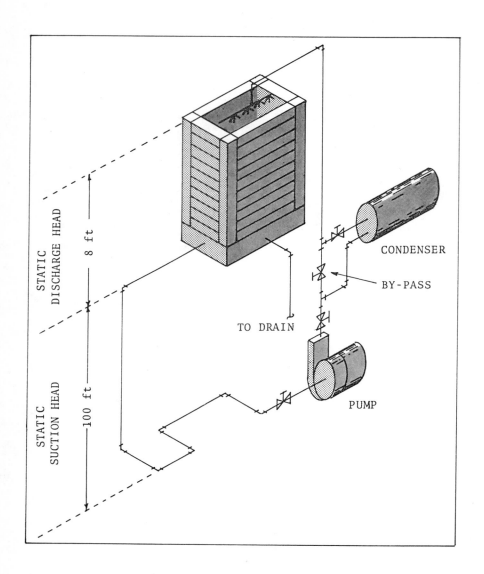

FIGURE 4.1

Table 4.4 shows a typical tower manufacturer's data sheet. It may be seen from this that depending upon the entering and leaving water temperatures and the design wet bulb conditions, a particular application may require 4, or even 5, gpm per ton across the tower. To balance the gpm flow over the condenser and tower, a by-pass should be piped in as shown in Figure 4.1.

STEP NUMBER 3; *Select a Preliminary Pipe Size.*

The samller the pipe the greater the friction; the greater the friction the greater the pumping horsepower. Figure 4.2 shows a pump-system balance curve to graphically explain the principle of pipe-pump system balance. The greater the pumping pressure the greater the quantity of water; as the flow increases, the pump curve drops. The system piping curve rises as the pressure across it is increased. The point at which both curves intersect is the balance point at which the system will actually operate. Figure 4.3 shows a friction chart for selecting pipe at the most economical balance point based on practical levels of water velocity.

PUMP SYSTEM BALANCE

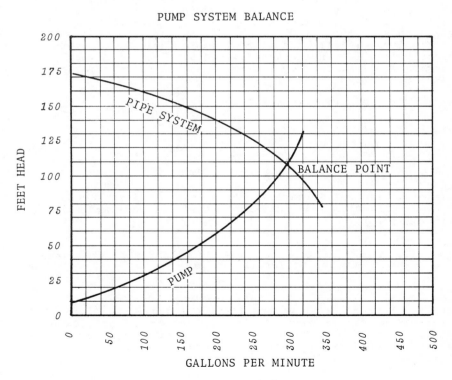

FIGURE 4.2

Knowing the required gpm from Step 2, enter the left ordinate of the friction chart at 300 gpm and follow horizontally to the shaded area. At the first diagonal line to intersect the 300 gpm horizontal within the shaded area, read the pipe size at 4 in. Where the gpm does not intersect a pipe line within the shaded area use the diagonal immediately above. Dropping down vertically from the intersection in the shaded area, the abscissa at the bottom of the scale will show the pressure drop in pounds per square inch per 100 feet of pipe length to be 4 lbs/sq in. To convert lbs/sq in. to head in feet use the formula:

Ft of H_2O = lbs/sq in. x 2.309

Using the above formula,

4 lbs/sq in. x 2.309 = 9.236 ft of head per 100 ft of pipe.

STEP NUMBER 4, *Sum the Friction Loss in the Pipe, Fittings, and Valves.*

Having selected the pipe size and found the friction loss per 100 feet of pipe, find the friction loss (friction loss, head in feet, pressure drop, are all interchangeable) for the total number of fittings and feet of pipe. Assuming 5 gate valves, 16 ells and 4 tees, go to Table 4.3 to find the equivalent length of these fittings. Table 4.3 shows approximate allowances for fittings and valves in terms of equivalent straight feet of pipe .*

4 gate valves x 4.4 =   17.6

11 ells x 10   =   110.0

2 tees x 6.5   =   <u>13.0</u>

equivalent ft of pipe   140.6

The combined friction loss for pipe and fittings equals 260 ft of pipe plus 141 equivalent ft of fittings and valves, totalling 401 ft.

$$401 \text{ ft x } \frac{9.23 \text{ ft}}{100 \text{ ft equivalent length}} = \frac{401 \text{ x } 9.23}{100} = 37 \text{ ft of head.}$$

*

Complete Tables for valve and fitting loses may be found in various handbooks such as, *Trane, Carrier, ASHRAE, Cameron Hydraulics, Crane;* or for speedy up-to-date data, the *Pipe-O-Graph* - designed by the author - is available from Technical Guide Publications, Miami

TABLE 4.2

TYPICAL MANUFACTURER'S CONDENSER RATING CHART

Gpm	150	170	190	230	270	310
Pd *	11	14	21	29	39	50

*
    Pressure drop in feet of water

STEP NUMBER 5; *Select the Cooling Tower.*

From Table 4.4, at the design conditions given, model number 4437 will handle 100.4 tons at 3 gpm/ton

STEP NUMBER 6; *Find the Head Loss of All Connected System Equipment.*

The total equipment loss includes the head loss through the tower sprays and the pressure drop across the condensers; these data must be determined from the manufacturer's published spec sheets. From Table 4.2 the pressure drop at the condenser is 50 ft. From Table 4.4 the tower spray pump head is shown as 8 ft.

TABLE 4.3

RESISTANCE OF VALVES AND FITTINGS TO FLUID FLOW-
PRESSURE DROP IN EQUIVALENT FEET OF STRAIGHT PIPE

Nominal Pipe Size	Fully Open Globe Valve	Fully Open Gate Valve	Standard 90° Ell	Standard 45° Ell	Flow Thru Run Tee
1	30	1.1	2.5	1.3	1.6
1-1/4	38	1.5	3.5	1.8	2.3
1-1/2	45	1.8	4.0	2.1	2.6
2	58	2.3	5.1	2.6	3.5
2-1/2	70	2.6	6.0	3.3	4.0
3	85	3.2	7.5	4.0	5.0
4	118	4.4	10.0	5.2	6.5

SPEED-0-GRAPH

WATER PIPE SIZE SELECTOR—FRICTION CHART

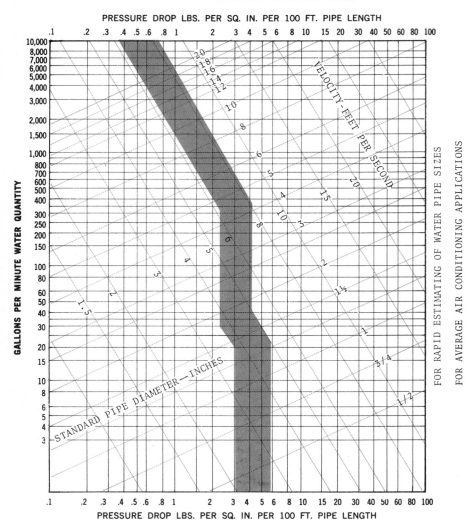

Enter the required gpm at the left ordinate and follow horizontally to the shaded area. At the first diagonal line to intersect the gpm within the shaded area, read the pipe size. Where the gpm does not intersect a pipe line with the shaded area, use the diagonal immediately above.

FIGURE 4.3

STEP NUMBER 7; *Find the Head Loss of the Vertical Lift.*

Pressure loss owing to the lifting of the column of water is only the height of the tower in that the height from the condenser to the tower is cancelled out by equal columns of water in the supply and return risers. In this example Table 4.4 shows that the vertical lift is 7 ft (dimension H).

STEP NUMBER 8; *Find the Total Pumping Head.*

From step number 4 the total combined loss for pipe and fittings is;   37 ft

From step number 6, Table 4.4 shows the loss for the tower sprays is;   8

From step 6, Table 4.2 shows the condenser loss is;   50

From step 7, the vertical lift is;   7

So, total pumping head required is;   102 ft

STEP NUMBER 9; *Select the Pump.*

From step 2 we determined 300 gpm, from step 8 we determined that the 300 gpm would have to be pumped against a total head of 102 ft. Now going to the typical manufacturer's curve sheet for a centrifugal pump as shown in Figure 4.4, we enter the left ordinate at the 110 ft line and go horizontally until we intersect the 300 gpm vertical line between the 10 hp and 15 hp broken lines. As in the case with the friction chart for pipe, always select the highest value when the intersection is between two points. The correct pump to handle this job would therefore be a 15 hp 2" x 3". To double check the selection apply the formula;

$$Bhp = \frac{gpm \times ft \ head}{2800} = \frac{300 \times 102}{2800} = 10.9 \ bhp$$

This checks with the selection above. Note; this formula is based on clear water and assuming the efficiency of the pump is 70%. Where the actual efficiency is known use the formula;

$$Bhp = \frac{gpm \times ft \ head}{3960 \times pump \ eff}$$

TYPICAL PUMP MANUFACTURER'S CURVE SHEET

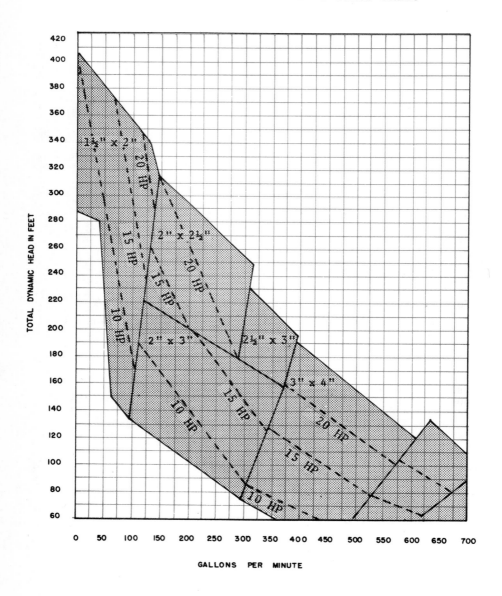

FIGURE 4.4

TABLE 4.4

TYPICAL TOWER MANUFACTURER'S RATING SHEET

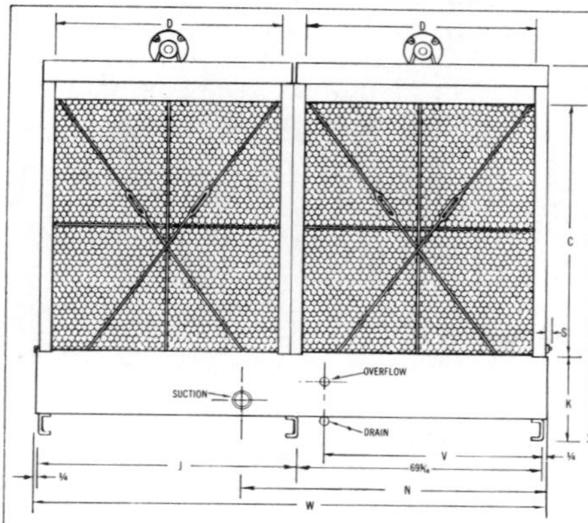

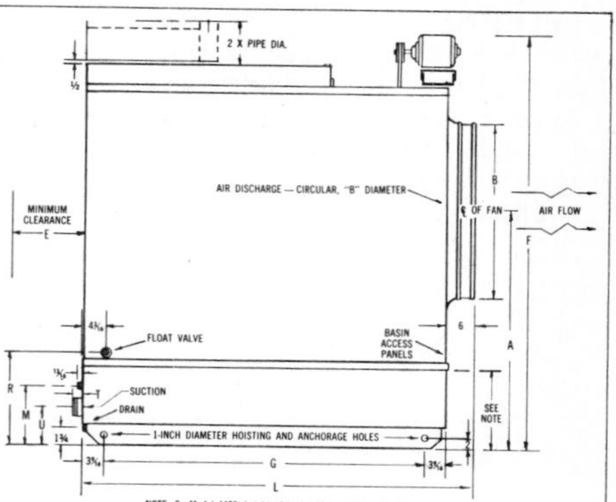

NOTE: On Model 4439, height of basin sides and front is 29"

MODELS 4435 THRU 4439—DIMENSIONAL DATA

Tower No.	SHIPPING DIMENSIONS W	L	H	DIMENSIONS IN INCHES A	B	C	D	E	F	G	J	K	M	N	P	R	S	T	U	V
4435	140¾	93	73¾	43 1/16	41⅞	49 3/16	66	72	83½	80	70 15/16	17	13¾	88⅜	32 15/16	18⅞	1	2	7¾	52⅜
4437	140¾	105	85¾	52 1/16	47⅞	61 3/16	66	84	95½	92	70 15/16	17	13¾	88⅜	32 15/16	18⅞	1	2	7¾	52⅜
4439	140¾	105	97¾	54 9/16	47⅞	73 3/16	66	84	109⅞	92	70 15/16	17	13¾	88⅜	32 15/16	18⅞	1	2	5	52⅜

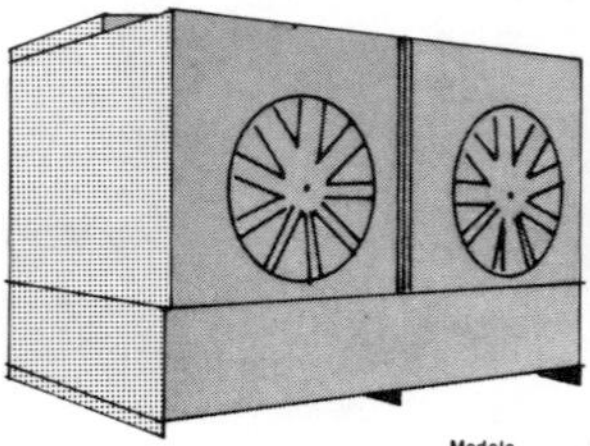

Models 4435-4437-4439

Tower No.	PIPING LAYOUT DATA—INCHES: Hot Water Inlet By Others	Cold Water Outlet	Over-Flow and Drain	Float Valve	Pump Head in Feet	MOTOR: HP	RPM	Approx. Shipping Weight** Less Motor	Approx. Motor Weight	Max. Oper-ating Weight
4435	4	6	2	¾	7.0	*1½	1750	3990	150	8670
4437	4	6	2	¾	8.0	*2	1750	4140	150	9085
4439	6	6	2	¾	9.0	*3	1750	4870	220	10660

*Two motors required, each at hp specified. **Freight classification: see specifications.
Note—Piping layout must provide bleed-off to prevent buildup of scale-forming solids. (See Installation Instructions.)

PERFORMANCE DATA

Tons of Refrigeration When Circulating 3 GPM Per Ton — Based on Dissipating 250 BTU Per Ton Per Minute

HOT WATER	90°	95°	90°	87°	95°	92°	96°	97°	95°	95°	97°	95°	96°	96°	97°
COLD WATER	80°	85°	80°	77°	85°	82°	86°	87°	85°	85°	87°	85°	86°	86°	87°
WET BULB	65°	70°	70°	70°	72°	72°	73°	75°	75°	77°	78°	78°	79°	80°	80°
4435	147.6	165.0	113.0	83.6	150.0	118.0	152.2	146.6	125.8	108.4	122.6	100.0	102.0	92.4	104.0
4437	177.0	198.0	135.4	100.4	179.2	142.0	182.8	176.6	150.6	130.2	147.0	120.0	122.4	111.2	124.4
4439	221.0	247.8	169.6	125.6	224.0	177.2	228.2	221.8	188.2	162.8	183.6	150.0	152.8	139.2	155.8

Tons of Refrigeration When Circulating 4 GPM Per Ton — Based on Dissipating 250 BTU Per Ton Per Minute

HOT WATER	82½°	85°	89°	89½°	95°	93½°	95°	95°	97°
COLD WATER	75°	77½°	81½°	82°	87½°	86°	87½°	87½°	89½°
WET BULB	65°	70°	70°	72°	78°	78°	79°	80°	80°
4435	97.2	85.0	125.4	114.2	123.0	104.4	114.0	105.6	128.0
4437	116.6	101.6	150.8	137.0	147.8	128.2	137.6	126.0	153.0
4439	145.8	127.2	188.4	171.2	184.4	160.2	159.4	157.8	191.8

Courtesy The Marley Company

STEP NUMBER 10; *Find the Bleed-off Rate.*

The required bleed-off, or blow-down, is usually based on ppm (parts per million) of solids in the circulating water and should be determined in actual operation by chemical analysis. As the tower water evaporates, the solids concentration in the remaining water increases; high concentrations of solids will cause scale formation and corrosion in the heat exchanger. For purpose of system design, Table 4.5 gives a simple method of approximating the bleed-off rate based on the tower range. In this case the rate will be 0.0033.

TABLE 4.5

RATE OF BLEED-OFF OR BLOW-DOWN

Cooling Range °F	Bleed-off Factor
6	0.0015
7-1/2	0.0020
10	0.0033
15	0.0054
20	0.0075

STEP NUMBER 11; *Determine the Required Make-up Water in Gpm.*

Under normal conditions a tower will evaporate about 1% for each 10 F (5.5 C) of range and will have a drift loss of about 0.2%. The required make-up must equal the losses from blow-down, evaporation, and wind drift.

Blow-down; from Step 10................ .0033

Evaporation; from given conditions the
  range is 10 F...................... .0100

Drift loss; from above............... .0020

                    Total loss   0.0153

Total make-up water required for the tower is;

0.0153 x 300 gpm = 4.59 gpm

Summary;

Gpm required to circulate		300 gpm
Economical pipe size		4 in.
Cooling tower size	# 4437-	100 tons
Pump size	# 2" x 3"	15 hp
Bleed-off	0.0033 x 300 gpm	1 gpm
Water make-up requirements		4.59 gpm

DUCT DESIGN

Unlike tower and pump problems, there is an abundance of discussions about ductwork in many handbooks and manuals as well as in textbooks and pamphlets; and problems on duct design will most certainly appear in any exam for air conditioning master. However, because the approach to duct design is many sided, duct design problems on examinations may vary widely. For this reason it is recommended that the candidate ground himself thoroughly in duct layout and air moving principles.

In air conditioning work a duct system will either be a *conventional (low velocity)* system, or it will be a *high velocity* system. Conventional systems usually range between 800 to 2000 fpm and sometimes as high as 2500 fpm — industrial applications may even go to 2700 fpm. The controlling factor is usually noise generation, so residential design would call for the lowest velocity; stores, cafeterias and industrial applications having a greater noise tolerance could accept higher velocities. Table 4.6 lists some recommended duct velocities for low velocity systems.

High velocity systems range from 2000 fpm upwards with the main supply duct at 4000 fpm. Noise is not a controlling factor in this system, therefore, within practical limits the ducts can be sized in accordance with the fundamental mechanics of fluid flow; the velocities are reduced at the special high velocity terminal units. To the author's knowledge, no license examinations for master contractor have included problems of high velocity duct design and it is not recommended that the reader spend time in this area.

Conventional duct system design may be accomplish by any one of the following three methods:

1. Static regain

2. Equal friction or constant friction

3. Velocity reduction or assumed reduction

TABLE 4.6

SUGGESTED DUCT VELOCITIES FOR LOW VELOCITY SYSTEMS

Application	Main Trunk and Risers	Branch Duct and Small Risers	Returns
Residences	800	600	600
Concert halls	900	700	700
Apartments	1000	800	800
Hotel, Motel bedrooms	1200	1100	1000
Theatres, Schools	1300	1100	1000
Executive offices, Libraries	1500	1200	1200
Dining rooms	1800	1400	1200
General offices	2200	1400	1200
Stores	2200	1600	1300
Cafeterias	2300	1800	1300
Industrial buildings	2600	1800	1500

STATIC REGAIN

The basis of the *static regain* method is to design the duct so that the velocity pressure decreases and the static pressure increases; a "static regain" i.e., a conversion of velocity pressure into static pressure. Although this is the most accurate method and the simplest to balance, it involves lengthy and complicated design procedures. For this reason it is not practical for average low velocity jobs. Furthermore, the *static regain* method requires more ductwork and consequently is more costly to install. Like high velocity duct design, *static regain* problems have not been appearing on examinations for contractors' license, so the candidate need not expect this category of questions.

EQUAL FRICTION

This is the most popular duct design system, favored by engineers as well as by contractors. It may be used to size exhaust systems as well as supply and return air conditioning systems, and most probably the examinee will encounter a duct layout problem requiring the *equal friction method*.

This method employs the same friction loss per foot of length for the entire system and *allows the system velocity to be governed by the system static.* The recommended procedure is to set the initial friction in accordance with an acceptable system velocity — see Table 4.6 — and then size the entire system accordingly. For example, a 1,200 cfm system for an apartment would require a velocity of 1000 fpm (from Table 4.6). From the air friction chart, Figure 4.5, or from the Trane Ductulator or similar tool, the comparative friction per 100 ft of duct is 0.1 in. Working along the same friction chart, the entire main trunk would be sized at 0.1 in. wg. A complete example will be discussed later.

VELOCITY REDUCTION

The least accurate and least desirable method is the *velocity reduction,* sometimes called the *assumed* velocity method. However uncommon its employment, it will usually show up on an examination and it is recommended that the candidate understand its application. The procedure is to set the initial velocity in the same manner as for the *equal friction* method and then arbitrarily reduce the velocity down the duct run *allowing the system static to be governed by the velocity.* Each subsequent reduction in velocity is a purely arbitrary function depending solely on the judgement and experience of the designer. A complete example will be discussed later.

GENERAL REQUIREMENTS FOR THE DUCT LAYOUT

The first requirement for designing a duct system is a good working knowledge of the fundamentals of fluid flow; and the second requirement is the acquisition of the basic data charts and/or slide rule devices such as the Trane *Ductulator* or the Carrier *Duct Calculator.* Examination text writers usually work from the *Trane Air Conditioning Manual* or the *Carrier System Design Manual,* the candidate should own, and be familiar with both of these books.

Figure 4.5 shows a typical friction chart with lines for cfm, velocity fpm, friction loss and round duct dimensions; given any two factors, the other two may be determined. Figure 4.6 gives a chart for converting round duct sizes to rectangular equivalents deriving from the formula;

$$d = 1.265 \sqrt[5]{\frac{(ab)^3}{a+b}}$$

Once the required round duct dimension is determined from Figure 4.5 the equivalent rectangular size can be found from Figure 4.6. The Trane *Ductulator* will locate all of these data simultaneously— and in a more graphic manner.

Equivalent Lengths

The friction loss or pressure drop of ductwork is usually given in *inches of water per 100 feet of length*. Threfore, the total pressure loss of a duct run must always be computed on the basis of the actual measured length. For example, assuming the friction loss for a particular duct system is 0.07 in. wg and the run is 229 ft, the total drop then would be;

$$\frac{0.07 \text{ in. wg}}{100 \text{ ft}} \times 229 \text{ ft} = 0.16 \text{ in. wg}$$

If the same duct run were 75 ft, then;

$$\frac{0.07 \text{ in. wg}}{100 \text{ ft}} \times 75 \text{ ft} = 0.05 \text{ in. wg}$$

Elbows and fittings, offering a greater resistance to flow than straight duct runs, must naturally be taken into account. To simplify and standardize the procedure, elbows and fittings are calculated on a basis of *equivalent length* of straight duct run. So each elbow has a value of resistance — depending on the configuration of the elbow — of X number of equal straight duct. The additional *equivalent length* must be added to the total length of straight duct run for the total friction loss.

Actually, for average, simply laid out systems, the practical designer will know from experience how the elbows and fittings will affect the design; he rarely has to calculate elbow losses. For average jobs — up to about 5000 cfm per main duct — using rectangular radius elbows, or rectangular square elbows fitted with turning vanes, a correctly fabricated elbow will have an equivalent length of about 12 ft. With say 10 elbows in the system and a friction loss of 0.08 for the ductwork, the total drop for the elbows would be;

$$\frac{0.08 \times 120 \text{ ft}}{100 \text{ ft}} = 0.096 \text{ in. wg}$$

or, about the equivalent of one average, clean filter.

Sometimes, however, when space requirements are tight and a "standard elbow" cannot be used, the *equivalent length* could go much higher. Some contractor examinations will require the candidate to calculate the value of each elbow. Table 4.7 gives *equivalent length values* for some common rectangular radius elbows; smooth round elbows and rectan-

SPEED-0-GRAPH

FRICTION LOSS FOR ROUND DUCT

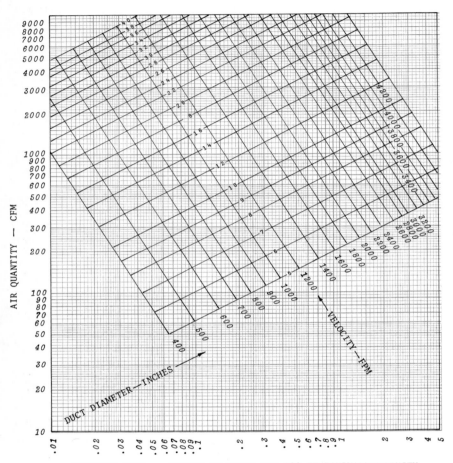

FRICTION LOSS — INCHES OF WATER PER 100 FT EQUIVALENT LENGTH

FIGURE 4.5

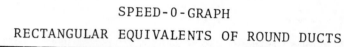

SPEED-0-GRAPH

RECTANGULAR EQUIVALENTS OF ROUND DUCTS

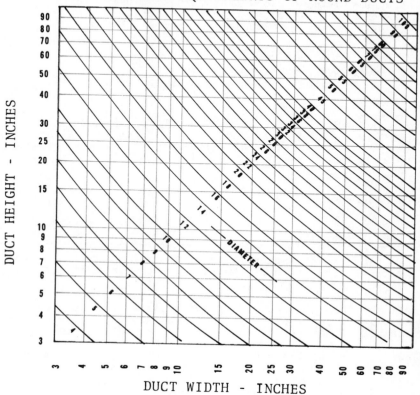

FIGURE 4.6

gular square elbows, fitted with turning vanes, have pretty much the same values. For other elbows and fittings see your *ASHRAE Handbook of Fundamentals, Trane Manual,* or *Carrier Handbook.* The terms used are:

R/W (radius ratio) = Elbow centerline radius ÷ width
H/W (aspect ratio) = Elbow height ÷ width
L/W (length ratio) = Value in feet per width

TABLE 4.7

EQUIVALENT LENGTHS OF STRAIGHT DUCT FOR
RECTANGULAR RADIUS ELBOWS WITH NO VANES

	L/W Ratio in Feet per Width *				
Radius Ratio R/W	Aspect Ratio= 0.5	Aspect Ratio= 1.0	Aspect Ratio= 2.0	Aspect Ratio= 3.0	Aspect Ratio= 4.0
0.75	13	18	25	30	35
1.00	9	11	13	15	17
1.25	6	7	8	9	10
1.50	3.5	4.5	5	5.5	6
1.75	2.75	3	3.5	3.75	4

* To find the equivalent feet of the elbow, multiply L/W ratio by the width of the elbow measured in inches ÷ 12.

Figure 4.7 shows how to find the elbow ratios. Different manuals may use different terminology, the candidate should be certain of terminology before using any charts or tables. An example of the use of radii in finding elbow losses;

Given: A rectangular radius elbow (no vanes) 24 in. wide and 12 in. high, 18 in. centerline radius.

Find: The equivalent length in feet of straight duct.

1. R/W = $\dfrac{18}{24}$ = 0.75 radius ratio

2. H/W = $\dfrac{12}{24}$ = 0.50 aspect ratio

3. L/W from Table 4.7 = 13

4. $\dfrac{13 \times 24}{12}$ = 26 ft equivalent length

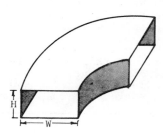

Aspect Ratio = Height / Width

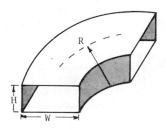

Radius Ratio = Centerline Radius / W

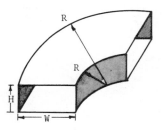

Curve Ratio CR = Throat Radius / Heel Radius

Standard Radius Elbow: An elbow with a curve ratio of 0.5 or a radius ratio of 1.5 is called a *Standard Radius Elbow* and is used wherever possible because it minimizes friction loss.

Turning Vanes, or splitters will reduce the aspect ratio.

FIGURE 4.7

FINDING ELBOW RATIOS

Elbow dimensions are usually such that the various ratios will not fall exactly on any table or chart and will therefore have to be interpolated. In the absence of any tables or charts, a good *rule of thumb* would be to allow 10 diameters (for round) or 10 widths (for rectangular) of equivalent length in feet of straight duct for each elbow.

Below is an actual problem taken from a recent examination for master mechanical contractor. In this particular examination the design method was *velocity reduction* but the solution is presented here in both *velocity reduction* and *equal friction,* so that the reader may make some comparisons.

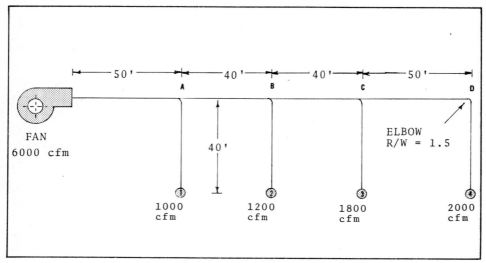

FIGURE 4.8

Given: 1. The layout in Figure 4.8
         2. Outlet loss is 0.05 in. wg
         3. Filters, coil and casing loss is 1.0 in. wg
         4. The architectural restrictions demand that the duct be kept to a 12 in. depth

Find: 1. Duct sizes for main and branch
        2. Friction loss for duct system
        3. Total system friction loss
        4. Horsepower of motor to drive the fan

Velocity Reduction

Step 1. Assume the initial velocity at 1300 fpm and from Figures 4.5 and 4.6 - or using a *Ductulator* - find the duct size; 70 x 12 (see item 4, Given) and the friction rate per 100 ft = 0.07.

Step 2. From the drawing in Figure 4.8, the measured length of duct from the fan to point A is 50 ft:

$$\frac{0.07}{1} \times \frac{50}{100} = 0.035$$

Therefore, the friction loss for the first run of duct is 0.035 in. wg

Step 3. Tabulate the information as shown in line 1 of Table 4.8. Notice that the last column is *longest run*. The static pressure requirement for the fan is determined by calculating the loss of the *longest run of duct work only*, including all the elbows and fittings in that run as well as the outlet loss at the end of the run. Therefore, the calculations for friction loss should be extended to column 8 for all losses occuring in the longest run of duct.

Step 4. At point A reduce the velocity to 1200 and figure the size and friction for 5000 cfm. Note that at point A, 1000 cfm was directed down branch #1. The reduction in velocity is purely arbitrary and is, therefore, a matter of judgement. By studying the relationships of velocity reduction to quantity reduction shown in Table 4.8, the reader will get a feel for the matter.

Step 5. Tabulate the loss for each run as shown.

Step 6 Figure all elbows separately in equivalent length. From the drawing, the elbow at point D has a *radius ratio* of 1.5, and the C-D duct dimension (line 4, Table 4.8) is 28x12; therefore, the aspect ratio is 12/28 = 0.43. Now, from Table 4.7 find the L/W = 3.5. So;

$$\frac{3.5 \times 28}{12} = 8.16 \text{ ft, say 8 ft.}$$

—add 8 ft. in equivalent length for the elbow at D.

Step 7. To find the fan horsepower, use the formula;

TABLE 4.8

DUCT CALCULATIONS

Section	Velocity	Cfm	Size	Length in feet	Friction Rate	Friction Loss	Longest Run inches wg
Fan to A	1300	6000	70x12	50	0.07	0.035	0.035
A to B	1200	5000	65x12	40	0.06	0.024	0.024
B to C	1100	3800	50x12	40	0.06	0.024	0.024
C to D	1000	2000	28x12	50	0.06	0.030	0.030
Elbow at D	——	——	28x12	8	0.06	0.005	0.005
Branch 1	850	1000	16x12	40	0.07	0.028	——
Branch 2	850	1200	19x12	40	0.06	0.024	——
Branch 3	850	1800	29x12	40	0.05	0.020	——
Branch 4	850	2000	32x12	40	0.05	0.020	0.020
					Friction loss		0.138

$$hp = \frac{Q \times fan\ Pt}{4000}$$

where,   Q = system cfm
          Pt = system total pressure in.
          .4000 = A constant based on 62.8% efficiency

Solution:

1. Table 4.8 gives the duct sizes

2. Duct friction loss = 0.138

3. Total system loss =

Ductwork	0.138
Outlet	0.050
Filters, coil and casing	1.000
Total inches of water	1.188

4. Horsepower to drive fan

$$hp = \frac{6000 \times 1.188}{4000} = 1.78$$

Equal Friction

Using the *equal friction* method for the same problem, and selecting an initial velocity of 1300 fpm as in the above method;

Step 1. The same as Step 1 in the *velocity reduction* method.

Step 2. Since the friction rate 0.07 will remain constant, columns 6 and 7 can be eliminated — see Table 4.9. Enter the measured length in column 5 of Table 4.9 and proceed to the next run of duct.

Step 3. Tabulate the size of each duct run based on 0.07 in. wg static.

Step 4. Total the number of feet in column 6 and find the duct friction loss;

Friction loss = $\frac{0.07}{100\ ft}$ x 230 ft = 0.161 in. wg

TABLE 4.9

DUCT CALCULATIONS

Section	Velocity	Cfm	Size	Length in feet	Longest run, ft
Fan to A	1300	6000	70x12	50	50
A to B	1250	5000	60x12	40	40
B to C	1200	3800	45x12	40	40
C to D	1000	2000	27x12	50	50
Elbow at D	——	——	27x12	10	10
Branch 1	850	1000	16x12	40	—
Branch 2	950	1200	17x12	40	—
Branch 3	950	1800	25x12	40	—
Branch 4	1000	2000	27x12	40	40

			Length of feet for longest run	230

Solution:

1. Table 4.9 gives the duct sizes

2. Duct friction loss = 0.161

3. Total system loss =

Ductwork	0.161
Outlet	0.050
Filters, coil and casing	1.000
Total inches of water	1.211

4. Horsepower to drive fan

$$hp = \frac{6000 \text{ x } 1.211}{4000} = 1.82$$

A comparison of the two methods shows that the *velocity reduction* system may consume an insignificantly lesser amount of power than the *equal friction* system, but the latter will have somewhat smaller ducts and therefore will usually be slightly lower in first cost.

Actually, the *equal friction* system provides easier balancing and greater reliability. In addition, it is simpler to design, and consequently, faster.

Many designers simply work with the *Ductulator* and ignore all other charts and tables for average jobs. Using the *equal friction* method and relying on experience, depending on the job, they may set the *Ductulator* at something like the following pressures; residences 0.08, offices 0.10, stores 0.15, industrial applications 0.25, and size the ductwork quickly. By keeping the aspect ratio close to 1.0 and the radius ratio close to 1.5, elbows and take-off fittings will have negligible friction losses.

BIBLIOGRAPHY

ASHRAE Handbook of Fundamentals, 1972, Chapter 24

Carrier Handbook of Air Conditioning System Design,
    McGraw-Hill, New York, 1965, Chapter 2

Carrier, W.H., Cherne, R.E., Grant, W.A., *Modern Air
    Conditioning, Heating and Ventilating,* 2nd ed.,
    Pitman, New York, 1959, Chapter 12

Trane Air Conditioning Manual, The Trane Co., 1965,
    Chapter 9

PSYCHROMETRIC PROBLEMS

One or more psychrometric problems are certain to appear on the examination. Usually, these are very simple questions to test your familiarity with the *psychrometric chart,* but occasionally, a psychrometric problem will be given that interconnects a load calculation with system selection; while not very difficult, it could be quite time consuming. It is suggested that the candidate becomes thoroughly familiar with the use of the psychrometric chart so that any problem can be solved directly and quickly.

Several versions of the psychrometric chart are available and albeit they may differ in varying degrees, all psych charts are based on the theory of *adiabatic saturation;* the psychrometric chart is simply a graphic representation of various air and water vapor mixtures. Consequently, irrespective of type of chart, the process lines on any chart will be similar — as shown in Figure 4.9.

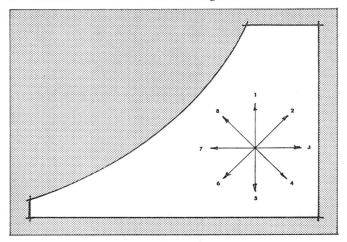

FIGURE 4.9

1.   add moisture
2.   add heat and moisture
3.   add heat
4.   add heat subtract moisture
5.   subtract moisture
6.   subtract heat and moisture
7.   subtract heat
8.   subtract heat add moisture

Knowing a particular chart well is an important advantage at an examination, for this reason it is suggested that the candidate study Chapter 6 of the *Trane Air Conditioning Manual* and Chapter 8 of the *Carrier Handbook of Air Conditioning System Design,* and decide with which psych chart he feels more comfortable. Having done so, he should work enough problems with that particular chart to develop speed and facility, and carry a pad into the examination room. Of course, there is the ASHRAE chart and several other charts published by equipment manufacturers, but there are variations in the construction of these charts. The major differences are in the amount of information one can extract from them, the speed with which they can be used, and the accuracy of the information. For the purposes set forth in this book, Carrier and Trane provide the best study discussion in conjunction with their psychrometric charts; the examples shown below are modifications of the Carrier method and Carrier psych chart.

Exactitude cannot be achieved, and should not be attempted, on this type of examination problem. Interpolations on the chart are constantly required, and the thickness of a pencil point could make a sizable difference on a small chart (8-1/2x11 format). *Sensible heat factor* slopes could easily be slightly out of angle causing an error at the point of *saturation intersection.* Finally, most psych charts are somewhat hard on one's eyes, and particularly for older persons, slight errors in reading may be expected. Most examination supervisors take these matters into consideration.

It is interesting to note that the Florida State Examination for Registered Professional Engineer requires that candidates taking the examination must solve the psychrometric problem portion by calculation rather than psych chart. This requires lengthy computations of great refinement, and an early mathematical error would naturally be compounded in the end result.

Psychrometrics require the use of a number of equations, and again, the wording and symbols of these equations may appear different in different applications. The need is to understand them and to apply them — the simpler the formulation the better; and the easier to read the better. There may even be several different formulas for one problem. The author has selected those with which he feels most comfortable; other formulas may be equally good — or better.

Figure 4.17 shows the Carrier psychrometric chart

BASIC PSYCHROMETRIC EQUATIONS

$$\text{Sensible heat} = \text{cfm} \times 1.08 \times \Delta t = \text{Btu/hr} \qquad (1)$$

$$\text{Latent heat} = \text{cfm} \times 0.68 \times \Delta \text{gr/lb} = \text{Btu/hr} \qquad (2)$$

$$\text{Total heat} = \text{cfm} \times 4.5 \times \Delta h = \text{Btu/hr} \qquad (3)$$

$$\text{Cfm supply to space} = \frac{\text{Room sensible heat}}{1.08 \ (\text{room db - leaving db})} = \text{cfm} \qquad (4)$$

$$\text{Entering db} = \text{Room db} + \frac{\%\text{OA}}{100} \ (\text{OAdb - room db}) = \text{Edb} \qquad (5)$$

$$\text{Leaving db} = \text{SI} + \text{BF} \ (\text{room db - SI}) = \text{Ldb} \qquad (6)$$

$$\text{Sensible heat factor} = \frac{\text{Room sensible Btu/hr}}{\text{Room total Btu/hr}} = \text{SHF} \qquad (7)$$

$$\text{Moisture rejection} = \frac{\text{cfm} \times 4.5 \times \Delta \text{gr/lb}}{7000} \div 8.3 = \text{gallons/hr} \qquad (8)$$

where, Δgr/lb = grains/lb OA - grains/lb room
  7000 = grains per pound of water
  8.3 = pounds of water per gallon

Other derivations of constants may be found on page 1-151 of the Carrier *Handbook*.

Symbols and Abbreviations

Δt	temperature difference
gr/lb	grains of moisture per pound of dry air
Δgr/lb	difference in gr/lb
SHF	sensible heat factor
BF	bypass factor
db	dry bulb
wb	wet bulb
h	enthalpy — total heat
Δh	enthalpy difference
Ldb	leaving dry bulb °F
Lwb	leaving wet bulb °F
Ewb	entering wet bulb °F
Edb	entering dry bulb °F
OA	outside air — design conditions
SI	saturation intersection
ADP	apparatus dewpoint

The major scales and process lines of a Carrier chart are shown in Figure 4.10. The *reference circle* on the Carrier chart is at 80 db and 50% rh. On the Trane chart the reference circle is at 78 db and 50% rh.

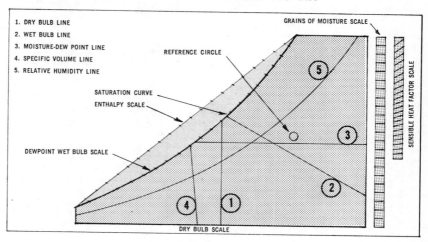

FIGURE 4.10

The example below is taken from an actual examination for Master General Mechanical - each step should be followed on a standard psychrometric chart.

Given:
1. Outdoor design   91°F db   79°F wb
2. Inside design   78°F db   67°F wb
3. Room sensible   100,000 Btu/hr
4. Room latent   17,000 Btu/hr
5. Outside air   1,000 cfm

Find:
1. Outdoor air load
2. Total coil load
3. Sensible heat factor
4. Leaving conditions at the coil
5. Supply air cfm
6. Entering conditions at the coil
7. Condensate rejection

Solution:

Step 1.a   Using equation 1;

1000 x 1.08 (91°F — 78°F) = 14,040 Btu/hr sensible OA

Step 1.b   Using equation 2;

1000 x 0.68 (131 gr/lb — 81 gr/lb) = 34,000 Btu/hr
                                          latent OA

Step 1.c   Using equation 3;

1000 x 4.5 (42.5*h* — 31.5*h*)= 49,500 Btu/hr total heat OA

Note: The above is a good example of the inevitable
inaccuracies that turn up in psychrometric solutions.   Where-
as Equation 3 gives 49,500 Btu/hr total heat, adding the
sensible and latent (from a and b) gives 48,040.   *Do not
spend time in an examination trying to reconcile these small
margins of error.*

Step 2.   From the conditions given (calculated from the
      heat load) add the room load to the outside air load;

room sensible	100,000
room latent	17,000
outside air	49,500
Total coil load =	166,500 Btu/hr

Step 3.   Using equation 7;

$$SHF = \frac{RSH}{RTH} = \frac{100,000}{117,000} = 0.85 \; SHF$$

Step 4.a.   Draw a straight line from the *reference
      circle* to 0.85 on the sensible heat factor scale.
      This is the *reference line.*

Step 4.b.   Using a pair of triangles, draw a straight
      line from the room air condition (78°F db, 67°F wb)
      parallel with the reference line, and intersect with
      the saturation curve.   This is the *SHF line.*   The
      point at which this line intersects the saturation
      curve, 59.5, is the *saturation intersection* (SI) see
      Figure 4.11

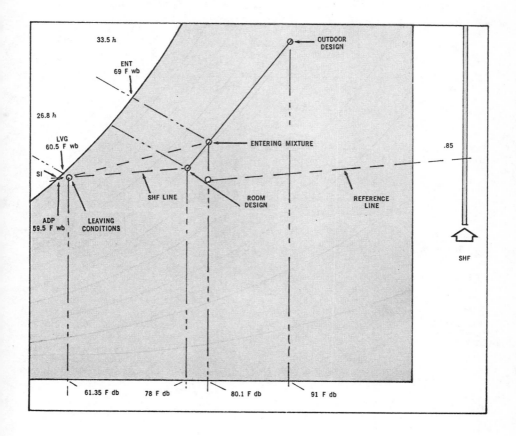

FIGURE 4.11

Step 4.c   Using Equation 6, BF must be assumed.   BF is
a *bypass factor*, and indicates the amount of air
passing through the coil that does not come in direct
contact with the fins or tubes. Typical bypass factors and a guide to selection may be found on page
1-127 of the *Carrier Manual*. For this example a BF
of 0.1 is assumed. So,

   59.5 + 0.1 (78 − 59.5) = 61.35 leaving dry bulb

Step 4.d Locate the leaving dry bulb, 61.35, at the intersection on the SHF line and read the leaving wet bulb temperature from the psych chart at 60.5.

Step 5. The cfm supply air to the space is found by Equation 4;

$$cfm = \frac{room\ sensible\ heat}{1.08\ (room\ db\ -\ leaving\ db)}$$

substituting in Equation 4,

$$\frac{100,000}{1.08\ (78\ -\ 61.35)} = 5561\ cfm$$

Step 6.a Determine the entering conditions by first finding the percent of outside air;

$$\%OA = \frac{1000}{6180}\ x\ 100 = 16.2\%$$

Step 6.b Using this percentage, calculate the entering dry bulb temperature from Equation 5,

$$Edb = 78 + \frac{16.2}{100}\ (91 - 78) = 78 + 0.162\ x\ 13 = 80.1°F\ edb$$

A good method of checking the accuracy of the calculations thus far, is to determine the total load on the coil using Equation 3;

Total heat = cfm x 4.5 x Δh

= 5561 x 4.5 x (33.5 − 26.8)

= 167,664 Btu/hr

This is slightly off from the total coil load found by addition in Step 2, but it is close enough.

Step 7   To find the condensate rejection use Equation 8;

$$\frac{5561\ x\ 4.5\ x\ 50}{7000} \div 8.3 = 21.53\ gal/hr$$

In practice, many designers develop their own forms and methods for figuring heat loads and solving psychrometric problems. For instance, some engineers do not use the bypass factor in their calculations but use a correction factor afterwards. Sometimes the leaving condition is not found by formula but arbitrarily placed on the SHF line at about 95% relative humidity. To determine the cfm supply air to the space, one large engineering office uses a simple formula: cfm = room sensible heat x 0.058. This formula is derived from a 16° Δt and would only be used for standard packaged unit applications. This is all to say that it is not common practice for contractors to figure heat loads or psychrometry to eight decimal accuracy. Usually, the examination writers are not interested in a candidate's experience at precision engineering, but rather in his ability to *understand* the nature of the problem and to come up with a *reasonable* solution.

REHEAT

There are many processes in industrial applications where the latent load is too high with respect to the total load and the solution can not be met in the simple manner outlined above, and plotted in Figure 4.11. This may also occur in high latent comfort cooling applications such as dance halls, skating rinks, bowling alleys, etc. In such cases the room design conditions can be altered somewhat and a low bypass factor coil can be selected. Under some circumstances this change in approach may solve the problem of an unacceptably low apparatus dew point. But in most cases the solution will require reheat. This means that the air — after leaving the cooling coil — must be reheated and the sensible heat factor changed to a more tolerable proportion. The procedure for determining the quantity of reheat follows.

Given:	1. Outdoor design	91°F db 79°F wb
	2. Inside design	78°F db 67°F wb
	3. Room sensible	80,000 Btu/hr
	4. Room latent	60,000 Btu/hr
	5. Outside air	1,000 cfm
Find:	1. Outdoor air load	
	2. Total coil load (before reheat)	
	3. Sensible heat factor	
	4. Reheat load	
	5. Total coil load (after reheat)	
	6. Leaving conditions at the coil	
	7. Supply air cfm	
	8. Entering conditions at the coil	
	9. Supply air temp of reheated air	

Solution:

Step 1.a Using Equation 1; Sensible heat = cfm x 1.08
x Δt = Btu/hr

1,000 x 1.08 (91°F -- 78°F) = 14,040 Btu/hr sensible OA

Step 1.b Using Equation 2; Latent heat = cfm x 0.68 x
Δgr/1b = Btu/hr

1,000 x 0.68 (131 gr/1b −81 gr/1b) = 34,000 Btu/hr
Latent OA

Step 1.c Using Equation 3; cfm x 4.5 Δh = Btu/hr

1,000 x 4.5 (42.5h − 31.5h) = 49,500 total heat OA

Step 2.   From the conditions given (calculated from
the heat load) add the room load to the outside air
load;

room sensible	80,000
room latent	60,000
outside air	49,500
	189,500 Btu/hr Total coil load

Step 3.   Using Equation 7;

$$\text{sensible heat factor} = \frac{\text{room sensible heat}}{\text{room total heat}} = \text{SHF}$$

$$\frac{80,000-}{140,000} = 0.57$$

Anyone accustomed to working at the design conditions
given above would know from experience that a SHF of 0.57
would result in a SHF slope that would either not intersect
the saturation curve, or would intersect at an unacceptably
low ADP. Now, by plotting the reference line, and parallel-
ing the SHF line, the *saturation intersection* point falls at
about 41°F. Since the leaving air will usually be within
90% of saturation, the ADP in this case would certainly be
too low.

From page 1-146 in the *Carrier Manual* a SHF of 0.65
would result in an ADP of about 51; this is acceptable.
Assuming a new SHF of 0.65 and scribing a new reference
line, the new *saturation intersection* point falls at about
53°F. These steps may be followed on the skeleton psych

chart in Figure 4.12.

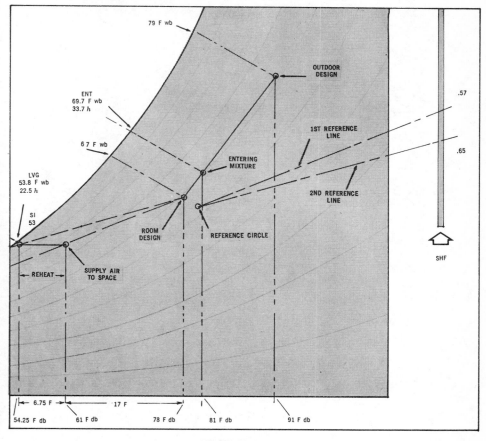

FIGURE 4.12

Step 4. To find the room sensible heat to be added by the reheat to the system it is first necessary to find the new RSH.

Since SHF = $\dfrac{RSH}{RSH + RLH}$ (Equation 7), then RSH = $\dfrac{SHF}{1 - SHF}$ x

RLH = $\dfrac{.65}{1 - .65}$ x 60,000 = $\dfrac{.65}{.35}$ x 60,000 = 111,428 BTU/hr

```
   111,428  Btu/hr (total RSH)
--  80,000  Btu/hr (RSH available from room load)
    31,428  Btu/hr (additional RSH required for reheat)
```

Step 5. The new total coil load is;

```
   111,428  RSH
    60,000  RLH
    49,500  OAH
   220,928  new total coil load
```

Step 6. Find the leaving dry bulb from Equation 6;

leaving dry bulb = SI + BF (room db − SI) = Ldb

= 53 + 0.05 (78 − 53) = 54.25°F leaving dry bulb.
Because of the high latent heat load, a bypass factor
of 0.05 was selected.

Step 7. Using Equation 4, the supply cfm may now be
calculated;

$$cfm = \frac{room\ sensible\ heat}{1.08\ (room\ db - leaving\ dry\ bulb)}$$

$$= \frac{80,000 + 31,428}{1.08\ (78 - 54.25)} = 4344\ cfm$$

Step 8. The entering conditions may be found using
Equation 5;

$$entering\ db = room\ db + \frac{\%OA}{100}\ (OAdb - room\ db) = Edb$$

$$= 78 + \frac{28}{100}\ (91 - 78\ ) = 81°F\ entering\ dry\ bulb$$

By placing the entering and leaving dry bulb on the
psych chart, the wet bulb conditions may be found.

Step 9. To determine the supply air temperature to
the space, divide the initial room sensible heat
(i.e., before the reheat load was added) by 1.08
x cfm and subtract from the room dry bulb. In
equation that will read;

$$supply\ air\ db = room\ db - \frac{RSH}{1.08\ x\ cfm}$$

$$= 78 - \frac{80,000}{1.08 \times 4344} = 61°F \quad \text{supply air dry bulb.}$$

The above 9 steps should be carefully followed on a psychrometric chart as illustrated in Figure 4.12.

As a check figure, the supply air db should be equal to Ldb plus the reheat Btu/hr divided by 1.08 x cfm, therefore:

$$54.25 + \frac{31,428}{1.08 \times 4344} = 61 \text{ F}$$

The grand total heat may also be checked graphically from the psych chart by applying Equation 3;

— from the chart, the entering enthalpy reads 33.7 and the leaving enthalpy reads 22.5

— so, 4344 x 4.5 x 11.2 = 218,937 Btu/hr GTH. This checks out closely with 220,928 from step 5.

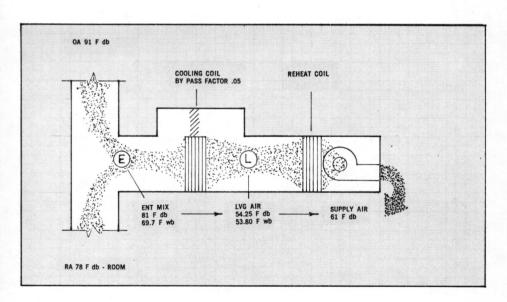

FIGURE 4.13

Figure 4.13, illustrates the reheating process by the use of a flow chart. Similar flow charts may be constructed for other processes as an aid to visualizing actual coil conditions where a thermometer may be inserted.

STANDARD AIR

All of the preceeding discussion is based on *standard air*, a term commonly used in designing air systems and rating air moving equipment. Standard air is established at 0.075 lbs per cu ft at 70°F and 29.92 in. hg barometric level. Whenever specificed air volumes are given at an altitute other than sea level or temperatures other than 70°F, correction factors must be used. But for normal air conditioning the correction factors need only be applied to systems designed to operate above 125°F, below 30°F, and altitudes above 2000 ft.

Figures 4.14 and 4.15 give the correction factors for altitude and temperature for air other than "standard." When a system operates at a temperature *and* altitude outside the standard range, the correction factors for both are multiplied together.

Example:

An air distribution system is specified for 60,000 cfm, at an altitude of 4500 ft and 120°F, what is the actual air quantity?

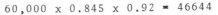

60,000 x 0.845 x 0.92 = 46644

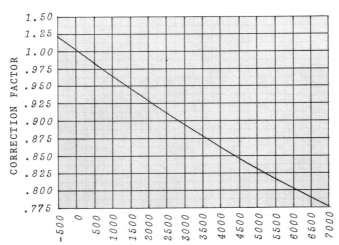

ALTITUDE — FFET ABOVE SEA LEVEL

FIGURE 4.14

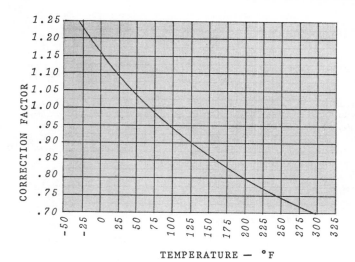

TEMPERATURE — °F

FIGURE 4.15

EVAPORATIVE AIR COOLING

Any rational analysis of the conditions that can be achieved by evaporative cooling requires a precise knowledge of the coincident values of local wet-and dry-bulb temperatures as well as the number of hours the wet-bulb temperature may be expected to go above "normal." For this reason the sizing of evaporative cooling equipment is usually made by rule-of-thumb method rather than any detailed calculations of building heat gain and Btu/hr values of equipment.

The most common rule-of-thumb is to divide the wet-bulb depression by 10 and set the quotient as the number of minutes required for each air change. For example:

Given: A centrifugal casting plant in New York City. Summer design conditions are 95°F db and 75°F wb. The plant is 80 ft long by 50 ft wide by 20 ft high.

Find: The size *evaporative air cooler* to achieve optimum conditions.

Solution:

The design conditions show the wet bulb depression to be 20° so, $\frac{20}{10}$ = 2 minute per air change.

$$\text{cfm supply air} = \frac{80 \times 50 \times 20}{2} = 40,000 \text{ cfm}$$

The evaporative air cooler selected from this example is one of 40,000 cfm capacity; or two units, 20,000 cfm each.

Occasionally, however, a problem may appear in an examination requiring a psychrometric solution for the selection of an evaporative air cooler; the candidate should be familiar with the psychrometrics of evaporative cooling processes. A graphic presentation of the following example is shown in Figure 4.16.

Given: 1. A centrifugal casting plant in New York City. Summer design conditions are 95°F db and 75°F wb.
2. Calculated;

sensible heat load	432,000	Btu/hr
latent heat load	232,615	Btu/hr
Total heat load	664,615	Btu/hr

3. Assume a 10°F temperature rise for the room.
4. Saturation efficiency of cooler 80%

Find: 1. Required air quantity
2. Leaving air condition at the evaporative cooler
3. Resultant room conditions

Solution:

Step 1.   From Equation 4

$$Q = \frac{432,000}{1.08 \times 10} = 40,000 \text{ cfm}$$

Step 2.a   First find the entering condition. Since an evaporative cooling system uses 100% outside air, the entering conditions in this case will be 95°F db and 75°F wb.

Step 2.b   The leaving conditions will fall somewhere along the wet bulb line of the entering air, close to saturation. The exact point along this line is determined by the saturation efficiency of the cooler. Evaporative coolers will usually operate between 75-95% saturation efficiency. In the conditions

given above, the saturation efficiency of
the cooler is 80%. This means that it will
cool the air the number of degrees equal to
80% of its original wet bulb depression.

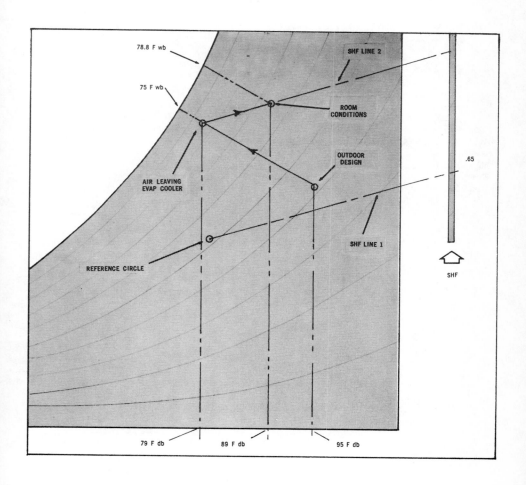

FIGURE 4.16

This relationship may be expressed as follows:

$$\text{Saturation efficiency} = \frac{db_1 - db_2}{db_1 - wb}$$

where;

db_1 = entering dry bulb temperature

db_2 = leaving dry bulb temperature

wb = entering/leaving wet bulb temperature

Therefore, the leaving dry bulb temperature (along the line of constant wet bulb temperature) is equal to the wet bulb depression x 80%.

Wet bulb depression = 95-75 = 20 deg
Temperature drop through the air cooler = 20 x 80% = 16 deg
Leaving dry bulb temperature = 95-16 = 79 deg

This solution may be expressed as follows:

$$DB_2 = DB_1 - E (DB_1 - WB)$$

$$= 95 - 0.8 \ (95 - 75) = 79 \ F$$

where;

E = saturation efficiency

Step 3.a Find the sensible heat factor;

$$SHF = \frac{432,000}{664,615} = 0.65$$

Step 3.b Draw the SHF slope of 0.65 from the leaving air condition; and from the assumed 10°F temperature rise given above, locate the room condition at 89 db and 78.8 wb on the SHF line.

It will be obvious from Figure 4.16 that climatological areas having low wet bulb depressions (below 18°F) will have minimal cooling effects. A complete discussion on evaporative air cooling may be found in Chapter 39 of the *ASHRAE Handbook and Product Directory 1973, Systems.*

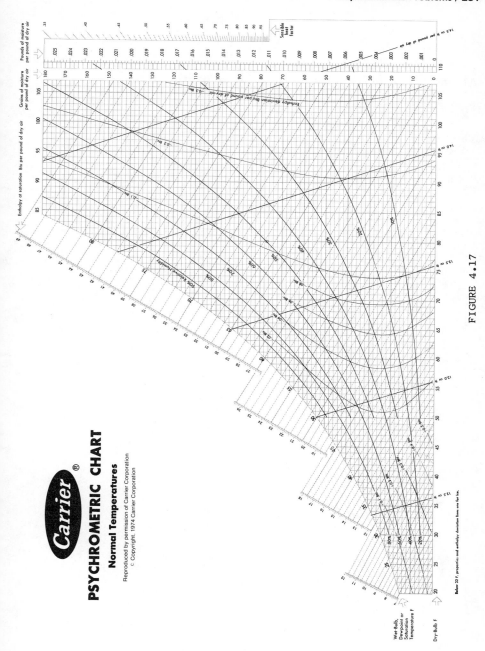

FIGURE 4.17

Carrier Psychrometric Chart

BIBLIOGRAPHY

ASHRAE Handbook of Fundamentals, 1972, Chapter 5

Carrier Handbook of Air Conditioning System Design, McGraw-Hill, New York, 1965, Chapter 8

Ramsey, M.A., *Tested Solutions to Design Problems in Air Conditioning and Refrigeration,* Industrial Press, New York, 1966

Trane Air Conditioning Manual, The Trane Co., 1965, Chapter 4

LOAD ESTIMATING

The basis for selecting the conditioning equipment is the *load estimate*; certainly, any error made in estimating the load will result in errors all along the line, so that the cost and performance of the final job would be disproportionate to the requirements. The *load estimate* is a problem that will appear on every exam and the candidate should be thoroughly versed in accurate load calculations.

Here again, different approaches may be made, and the final solution will always depend upon the judgement and experience of the designer. A complete discussion on heat gains may be found in three major sources; 1. ASHRAE Handbook of Fundamentals, 1972, Chapters 20, 21 and 22. 2. Trane Air Conditioning Manual, Chapter 3. 3. Carrier Handbook of Air Conditioning Systems Design, Part 1.

To the novice, heat gain calculations may seem complex and tedious, but once mastered they are quite simple. The designer who does these calculations regularly can perform the task with great speed and little effort; he knows exactly which tables to use and *where to find them*. He has all his material organized and uses forms and worksheets with which he has long been familiar.

In the examination room, the candidate cannot afford a wasted moment. He must already be familiar with a set of work sheets and forms, and he must have his reference books marked in such a manner that he can go quickly to the required tables and charts.

Heat loads, like psychrometrics, cannot be mastered by discussions, or from reading texts on how to do it. The only way to learn heat load estimating is to spend many quiet, undisturbed hours practicing...doing examples over and over, until the meaning of each step becomes crystal clear; until the "mystery" of the major equations unravel, and become keys to the doors through which the designer must pass to arrive at a solution. It is suggested that the reader first go through the following steps quickly, marking off those passages and equations which are unclear. Then going back to his notes, and with the aid of the Carrier Handbook, Trane Manual, or ASHRAE Handbook, work over each point in question carefully until it is thoroughly understood. Take the time to rule some work sheets and summary forms (or these may be purchased), and rework a few examples several times over. *It is important to learn to do them quickly.* The most common cause of failure in examination is running out of time.

Where to Find the Data

   The candidate should work from the reference book with
which he is most familiar and comfortable. The author's
preference is the *Carrier Design Handbook,* and for the purpose of illustrations, tables cited will be from that reference.

   A master index of major tables should be prepared as for
example the following from the *Carrier Handbook:*

	ITEM	TABLE	PAGE*
1.	Design weather data	1,2,3	10 to 19
2.	Inside design conditions	4,5,	20,22,23
3.	Glass loading		
	a. Direct solar heat gain	15	44 to 49
	b. Adjustable shade factors	16	52
	c. Fixed shading factors		55 to 58
	d. Transmission coefficients	33	76
4.	Solar and transmission		
	a. Equivalent temperature difference, walls	19	62
	b. Equivalent temperature difference, roofs	20	63
	c. Transmission coefficient U, walls	21 to 24	66 to 68
	d. Transmission coefficient U, partitions	25,26	69,70
	e. Transmission coefficient U, roofs	27,28	71,72
	f. Transmission coefficient U, ceilings and floors	29	73
	g. Transmission coefficient U, insulation, air space	31,32	75,76
5.	Internal heat		
	a. People	48	100
	b. Lights, storage factors	12	35
	c. Lights, direct heat gain	49	101
	d. Restaurant appliances	50,51	101,102
	e. Miscellaneous appliances	52	103
	f. Electric motors	53	105
6.	Outdoor air heat		
	a. Infiltration	41 to 44	90 to 96
	b. Ventilation	45	97

*   All page numbers are from Part 1 of the *Carrier Handbook
of Air Conditioning System Design.*

The above index includes all of the major tables necessary to compute the heat load; other tables, used infrequently, are omitted. Usually, the examination problems are simple
and will not require reference to all the tables indexed,
nevertheless, the candidate must know where to find the data
if he needs them.

A master index of the major tables is a prerequisite for
finding solutions to problems of heat load. Another requisite
is a proper load estimating form on which to gather the known
facts in an orderly way. A proper form will organize the
information in such a way as to achieve sequentiality and ease
of computation; it will also function as a check sheet to
make certain that none of the essential facts is left out of
the heat load. The Trane *Load Estimate Sheet* form number
26.04, and the Carrier *Cooling and Dehumidifying Estimate,*
form number E-20, are good examples of such forms. The
examinee should have a supply of either of these forms or
one used by his office to estimate heat loads.

USING THE FORMS

Figure 4.18 shows a *Cooling Load Estimate Summary*. This
form was designed by the author and is based on the Carrier
System, with some additions and some omissions. It is a
simplified form that is particularly good for exam work.
Figures 4.19 and 4.20 show the work sheets that are used to
prepare the information before it is transferred to the
Estimate Summary. These forms help the user organize his
data and arrive at a quick determination of the cooling load.
The summary sheet also has room for check figures; usually
these check figures will reveal an inconsistancy in the event
of a major error.

Storage factors, diversity factors, precooling and
stratification are some of the items that are not included.
The contractor's examination is not an engineer's examination
and the examinee is not expected to split decimal points. Of
those tests that have been scrutinized, none was found to
require a complicated heat load based on built-up equipment;
a package unit application is all that has been required up
to the present time. For this reason, it is felt that the
candidate should not put anymore into the exam than is expected of him. There simply isn't enough time. Practically,
a package unit job is quite limited in selection of number
of coils, capacity, partial load operation, humidity control
etc., and precise heat loads for this type of equipment are
not justified. The forms and methods recommended in the
example in this chapter should suffice for any contractor

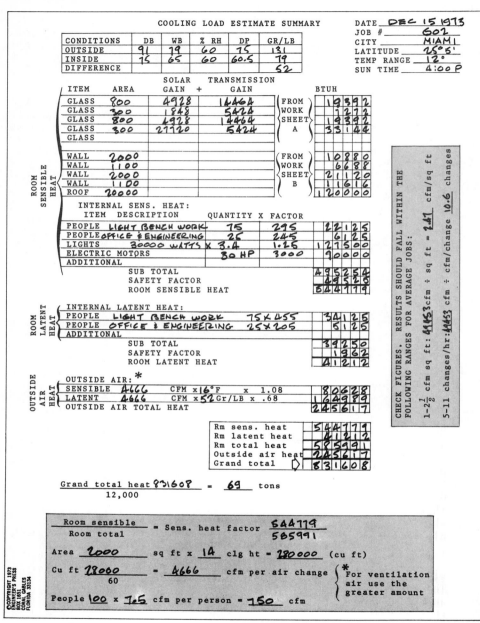

FIGURE 4.18

WORK SHEET A FOR GLASS LOAD

SUN TIME 4.00 PM

DIRECT SOLAR GAIN

ITEM	Col 1 AREA NET SQ FT x	Col 2 SHADE COEFFICIENT x	Col 3 SOLAR GAIN BTUH/ SQ FT =	Col 4 ACTUAL SOLAR GAIN
N.Glass	800	.56	11	4928
E.Glass	300	.56	11	1848
S.Glass	800	.56	11	4928
W.Glass	300	.56	165	27720
N.E.				
S.E.				
N.W.				
S.W.				

TRANSMISSION GAIN

ITEM	Col 5 AREA NET SQ FT x	Col 6 COEFFICIENT U	Col 7 TEMPERATURE x DIFFERENCE =	Col 8 TRANSMISSION GAIN
N.Glass	800	1.13	16	14464
E.Glass	300	1.13	16	5424
S.Glass	800	1.13	16	14464
W.Glass	300	1.13	16	5424
N.E.				
S.E.				
N.W.				
S.W.				

TOTAL GAIN FOR ALL GLASS

ITEM	Col 9 ACTUAL SOLAR GAIN (from Col 4) +	Col 10 TRANSMISSION GAIN (from Col 8) =	Col 11 TOTAL BTUH
N.Glass	4928	14464	19392
E.Glass	1848	5424	7272
S.Glass	4928	14464	19392
W.Glass	27720	5424	33144
N.E.			
S.E.			
N.W.			
S.W.			

FIGURE 4.19

WORK SHEET   B   FOR WALLS AND ROOF

SUN TIME   4:00 PM

SOLAR AND TRANSMISSION GAIN — WALLS AND ROOF

ITEM	Col 1 AREA NET SQ FT x	Col 2 COEFFICIENT U	Col 3 EQUIVALENT* x TEMP DIFF =	Col 4 TOTAL BTUH
WALL N	2000	.32	17	10880
WALL E	1100	.32	19	6688
WALL S	2000	.32	33	21120
WALL W	1100	.32	33	11616
WALL				
ROOF/SUN	20000	.12	50	120000
ROOF/SHADE				

TRANSMISSION GAIN — EXCEPT WALLS AND ROOF

ITEM	Col 5 AREA NET SQ FT x	Col 6 COEFFICIENT U	Col 7 TEMPERATURE x DIFFERENCE** =	Col 8 TOTAL BTUH
PARTITION				
PARTITION				
CEILING				
FLOOR				

 * Tables of equivalent temperature are usually presented
for one latitude, daily temperature range, temperature
difference and month of the year; Column 3 above is for
the *corrected* equivalent difference.

** The temperature difference for walls, ceiling, and partitions is the outside temperature minus the inside
temp. minus 5 F, when the space on the other side is
unconditioned. If the adjacent space is a boiler room
or kitchen then the temperature difference becomes;
outside temp. minus inside temp. + 20 F.

WORK SHEET   C   FOR FINDING MAXIMUM LOAD

ITEM	Col 1 SUN TIME 12:00	Col 2 2:00	Col 3 4:00
GLASS N	6275	6275	4928
GLASS E	2352	2352	1848
GLASS S	11648	8960	4928
GLASS W	2352	17808	27120
ROOF	55200	96000	120000
TOTAL	77827	131395	159424

FIGURE 4.20

examination offered anywhere.

BASIC HEAT LOAD EQUATIONS

External Sensible Heat Gain

1. Heat gain through glass: (See worksheet "A")

a. (Area sq ft)x (solar heat gain) x (shade coefficient) = Btuh (9)
    (Carrier Table 15, pp 44-49) (Carrier Table 16, p 52)

b. (Area)x (coefficient U) x(outdoor temp — indoor temp)= Btuh (10)
    (Carrier Table 33, p 76)

c. Solar gain (a) + transmission gain (b) = Total glass gain
                                                        Btuh (11)

2. Heat gain through walls: (See worksheet "B")

a. (Area sq ft)x (coefficient U) x (equivalent temp difference)=
                                                        Btuh (12)
        (Carrier Table 21-26,pp 66-70)
        (Carrier Table 19, p 62, with corrections)

3. Heat gain through roof: (See worksheet "B")

a. (Area sq ft)x (coefficient U) x (equivalent temp difference)=
                                                        Btuh (13)
        (Carrier Table 27, p 71)
        (Carrier Table 20, p 63, with corrections)

4. Heat gain through partitions: (See worksheet "B")

a. (Area sq ft)x (coefficient U) x (outside temp — inside temp)=
                                                        Btuh (14)
        (Carrier Table 25, p 69)   (minus 5 F)

    Note: When boiler room or kitchen adjacent to the
           partition wall, add 20 F to temp difference.

5. Heat gain through floor: (See worksheet "B")

a. (Area sq ft)x(coefficient U)x (outside temp — inside temp) =
                                                        Btuh (15)
        (Carrier Tables 29 and 30, pp 73, 74) (minus 5 F)

Internal Sensible Heat Gain

6. Heat gain from people:

    a. (Number of persons)x (sensible heat factor) = Btuh   (16)
          (Carrier Table 48, p 100)

7. Heat gain from lighting:

    a. Total incandescent wattage x 3.4 = Btuh   (17)

    b. Total fluorescent wattage x 3.4 x 1.25 = Btuh   (18)

8. Heat gain from electric motors:

    a. Under 3 horsepower; hp x 3600 = Btuh   (19)

    b. Three horsepower and up; hp x 3000 = Btuh   (20)

Internal Latent Heat Gain

9. Heat gain from people:

    a. (Number of persons)x (latent factor) = Btuh   (21)
          (Carrier Table 48, p 100)

Outside Air Heat

10. (Cfm)x(inside temp — outside temp)x(1.08)= Sensible Btuh (22)

11. (Cfm)x (grains per lb) x(0.68)= Latent Btuh   (23)

12. Quantity cfm OA =

    a. Cubical contents ÷ 60 minutes; or,   (24)

    b. Number of people x cfm per person
       — whichever is the greater. (Carrier Table 45, p. 97)

SEVEN EASY STEPS TO SOLVING THE HEAT LOAD

    The seven basic steps to finding the load and selecting
the equipment for any cooling job are;

1. Find the total glass load
2. Find the wall and roof load
3. Find the time of the maximum peak load
4. Calculate the internal sensible heat gains
5. Calculate the internal latent heat gains
6. Calculate the outside air heat
7. Solve the psychrometry of the load; that is, find the

SHF, ADP, air quantity across the coil, entering and leaving conditions, reheat, etc.

In an examination the design conditions may or may not be given. If not, they may be found from the appropriate Tables. Other conditions must be given; the materials of construction for roof or walls, glass areas, etc., cannot be assumed. The following example is taken from an actual South Florida examination for air conditioning contractors license and the solution is arrived at using the seven steps above:

Given: An aluminum window factory in the Miami area with:

1. Dimensions;

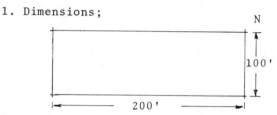

2. Roof; 2" gypsum slab on 1/2" gypsum board with 1" top deck insulation.
3. Ceiling; 14' high, suspended acoustic tile
4. Walls; 8" concrete block (lt wt agg), 3/8" gyp board inside.
5. Floor; on slab
6. People; 75 at benches and machines, 25 engineering and office.
7. Operating hours; 8:00 AM to 5:00 PM
8. Illumination; 30,000 watts fluorescent
9. Motors; 30 horsepower
10. Windows; uncoated, single, with light colored inside venetian blinds — no outside shading devices. 800 sq ft in north wall, 300 sq ft east, 800 sq ft south and 300 sq ft west.

Solution: Enter the design conditions in the top of the *Cooling Load Estimate Summary*, Figure 4.18. Design conditions may be established from Tables 1,2,3,4 and 5.

Step 1. From Carrier Table 15, the peak glass load is 4:00 PM, August. Enter the figures onto Work Sheet A, Figure 4.19. N. 11, E. 11, S. 11, W. 165. From Carrier Table 16, the shade coefficient is found to be .56. Note that Miami is 25 deg 5 min north latitude, and Table 15 20 and 30 deg; we have used 20 deg.

```
800 sq ft N x .56 x 11 =    4928
300 sq ft E x .56 x 11 =    1848
800 sq ft S x .56 x 11 =    4928          (9)
300 sq ft W x .56 x 165 = 27720
```

From Carrier Table 33, the coefficient U for single glass is 1.13. The temperature differ-ence was established at the design conditions.

```
800 x 1.13 x 16 =  14464
300 x 1.13 x 16 =   5424          (10)
800 x 1.13 x 16 =  14464
300 x 1.13 x 16 =   5424
```

Still using Work Sheet A, Figure 4.19, the total Btuhr for glass becomes;

```
N.  4928 + 14464 =  19392  Total Btuh
E.  1848 +  5424 =   7272  Total Btuh
S.  4928 + 14464 =  19392  Total Btuh
W. 27720 +  5424 =  33144  Total Btuh
```

These numbers may now be transferred to the Summary Sheet;

Step 2. Subtracting the glass area from each wall area gives the *net sq ft* of wall.

```
N. wall  200' x 14' = 2800 sq ft - 800 =  2000  net
E. wall  100' x 14' = 1400 sq ft - 300 =  1100  net
S. wall  200' x 14' = 2800 sq ft - 800 =  2000  net
W. wall  100' x 14' = 1400 sq ft - 300 =  1100  net
```

From Carrier Table 21, the coefficient U for 8 in. concrete block, light weight aggregate with 3/8 in. finish is .32. From Carrier Table 27, the coefficient U for the roof is .12. Enter these numbers on Work Sheet B, Figure 4.20.

From Carrier Table 19, the equivalent temper-ature difference for 4:00 PM (remember the maximum glass load was 4:00 PM) is; N. 10, E. 12, S. 26, W. 26. And from Carrier Table 20, the equivalent temperature difference for the roof is 43. Notice that these tables are based on a 15 deg temperature difference; 20 deg daily range, 24-hour operation and 40° N. latitude. Therefore, these figures must be corrected. From Carrier Table 20A for 12 deg range and 16 deg Δt, select a factor of +5 and add another 2 deg. for the difference in operating time and latitude. Work Sheet B will now read;

N. wall	2000	x .32	x 17	=	10880	Btuh	
E. wall	1100	x .32	x 19	=	6688	Btuh	
S. wall	2000	x .32	x 33	=	21120	Btuh	(12)
W. wall	1100	x .32	x 33	=	11616	Btuh	
Roof	20,000	x .12	x 50		120000	Btuh	(13)

Step 3. The experienced designer will usually know at a glance what the time of the maximum load will be by judging the peak temperature differentials through sunlit roof coincident with peak solar transmission through glass. The internal heat gain from people, lights, and electric motors will be constant between the specified hours of operation. A large exposure of west glass will almost always indicate a 4:00 PM maximum load, and at any rate — depending on the building configuration — the maximum load should fall between 12:00 PM and 4:00 PM. Work Sheet C, Figure 4.20 provides a comparison for finding the maximum load.

Without reviewing the actual computations for each sun time, Work Sheet C would look as follows:

	12:00	2:00	4:00
N.	6275	6275	4928
E.	2352	2352	1848
S.	11648	8960	4928
W.	2352	17808	27720
Roof	55200	96000	120000
TOTAL	77827	131395	159424

Obviously, a large exposure of east glass may well peak the load at 8:00 AM or 9:00 AM, in a particular building; and this points up the extreme importance of finding the time of maximum load for each exposure where multi zone design is considered.

In this particular example, the Examining Board asked the examinees to "find the summer load." Were that not so, a determination of the *maximum winter cooling load* would also have to be made. In fact, owing to the large exposure of south glass, the maximum temperature differential through the roof coincident with peak solar transmission through glass would occur at noon in December!

Step 4. Calculating the internal sensible heat gain, from Carrier Table 48;

75 people — light bench work x 295 =   22125
25 people — office and engineering x 245 = 6125   (16)

The lighting load is calculated from equation 18;

30,000 fluorescent watts x 3.4 x 1.25 =   127500

From equation 20 the heat gain from electric motors is;

30 horsepower x 3000 =   9000.   (20)

That completes the calculations for the room sensible heat, and when entered onto the *Cooling Load Estimate Summary* will come to a sub total of 495254 Btuh. Accepted practice calls for the addition of 10% to the sensible sub total to allow for air leakage through the duct seams, by passed air at the coil, supply duct heat gain, plus a factor for probable error in the estimate. So; 495254 + 49525 = 544779 Btuh Total room sensible heat.

Step 5. The internal latent heat gain is calculated from Carrier Table 48;

75 people — light bench work x 455 =   34125
25 people — office and engineering x 205 = 5125   (21)

Accepted practice calls for a 5% safety factor to be added to the room latent bringing the total room latent heat to 41212 Btuh.

Step 6. The quantity of outside air is found by equation 24, which in this case will be;

$$\frac{280,000}{60} \text{ cu ft} = 4666 \text{ cfm per air change} \qquad (24)$$

The sensible heat and latent heat of the outside air entering the room may be calculated from equations 22 and 23 respectively;

a. 4666 cfm x 16 F Δt x 1.08 = 80628 Btuh OA sensible (22)

b. 4666 cfm x 52 grains/lb x 0.68 = 164989 Btuh OA latent (23)

Step 7. The final step is to determine the cfm supply to the space and the necessary psychrometrics. These calculations have already been discussed in the earlier chapter, Psychrometrics. Briefly — applying the psychrometric equations — the basic solutions are:

a. Sensible heat factor = $\dfrac{544779}{585991}$ = .93 SHF (1)

b. Apparatus dewpoint, from Carrier Table 65 p.146

= 58 ADP

c. Leaving dry bulb = saturation intersection + BF (rm db — sat. inter.) = 58 + .4 (75 F - 58 F) = 64.8 leaving db (6)

This is assuming a bypass factor of .4 for a two row coil.

d. Cfm supply = $\dfrac{\text{room sensible heat}}{1.08\ (\text{room db - leaving db})}$

= $\dfrac{544779}{1.08 \times 75 - 64.8\ F}$ = 49453 cfm (4)

Comparing the check figure ratios on the side of the *Estimate Summary*:

$\dfrac{49453\ cfm}{20000\ sq\ ft}$ = 2.47 cfm per sq ft

$\dfrac{49453\ cfm}{4666\ OA}$ = 10.6 air changes per hour

Both of these check figures indicate a high sensible heat load but fall within the acceptable range for this application.

The grand total heat, incidentally, is;

Room sensible heat	=	544779
Room latent heat	=	41212
Outside air heat	=	245617

Grand total heat	831608 Btuh

$$\frac{831608}{12,000} = 69 \text{ tons}$$

SPECIAL CONSIDERATIONS

It was stated earlier that during an examination, time would not allow the candidate to provide more information in his answer than he is asked to give. Answers to problems should be limited to the scope of the question, and although the answer should be complete in every regard the candidate should not try to split decimal points or volunteer solutions beyond the requirements. This, however, brings up some matters of judgement that could be critical.

Bypass Factors

It would be proper and correct for the test writers to include a given bypass factor in the known conditions; however, the author has not seen any examinations where the bypass factor was given. Therefore, the bypass factor must be assumed — or, omitted completely.

In the example used above, a bypass factor of .4 was assumed and from equation 6 the leaving dry bulb was 64.8; and from equation 4 the cfm to the space was 49,525. This is a very high ratio of cfm per ton. Since this job is obviously a high sensible heat application with a large quantity of outside air, a coil with a low bypass factor should have been selected. A comparison of different bypass factors appears as follows:

.4 BF = 75 F − 58 F = 17.   17 x .4 + 58 =   64.8 LDB

$$\frac{544779}{1.08 \text{ x } 75 - 64.8} = 49453 \text{ cfm}$$

.3 BF = 75 F − 57 F = 17.   17 x .3 + 58 =   63.1 LDB

$$\frac{544779}{1.08 \text{ x } 75 - 63.1} = 45779 \text{ cfm}$$

$.2$ BF = 75 F — 58 F = 17.   17 x .2 + 57 =   61.4   LDB

$$\frac{544779}{1.08 \times 75 - 61.4} = 37090 \text{ cfm}$$

$.15$ BF = 75 F — 58 F = 17.   17 x .15 + 58 =   60.5   LDB

$$\frac{544779}{1.08 \times 75 - 60.5} = 34,787 \text{ cfm}$$

It may be seen by these comparisons that the air requirements for a .4 bypass factor coil and a .15 bypass factor
coil differ by 30% and the cfm to tonnage ratio changes from
717 cfm per ton to 504 cfm per ton. The reader should pay
particular attention to Tables 61 and 62, Carrier page 127,
and read the related discussion.

Illumination and the Lighting Load

The unit of illumination is the footcandle, fc. It
represents a surface of one square foot on which there is a
uniformly distributed flux of one lumen; one footcandle
equals one lumen per sq ft. Sometimes an architectural plan
is made part of an examination and the heat load must be
worked up from that plan. However infrequently, an architectural or interior designer's plan may show the lighting
load in footcandles. Table 4.10 gives Fluorescent Lamp
Data for general reference when plans omit wattage and give
either lamp sizes or footcandles.

Table 4.11 is a convenient reference to use when no wattage or illumination data are available and the designer must
assume a light load based on some kind of experience rate.

TABLE 4.10

FLUORESCENT LAMP DATA

SLIMLINE	T6		T8		T12	
Nominal length	42	64	72	96	72	96
Watts	25	38	37	50	56	74
Lumens per watt (max)	66	71	73	76	75	78

STRAIGHT RAPID START	430 MA		800 MA		1500 MA	
Nominal length	36	48	72	96	72	96
Watts	30	40	85	105	150	200
Lumens per watt (max)	65	75	68	80	76	82

HOT CATHODE					
Nominal length	18	24	36	48	60
Watts	15	20	30	40	91
Lumens per watt (max)	47	56	65	74	64

TABLE 4.11

LIGHTING LOAD CHECK FIGURES

APPLICATION	WATTS/ SQ FT
Apartments	1.00
Libraries	1.50
Restaurants	1.75
Offices	2.00
Banks	2.50
Dept. Stores	3.00

Usage Factors

Factory motors, beauty shop hair dryers, restaurant equipment, and other appliances will seldom operate simultaneously 100% of the time. And likewise, in the case of the example given above, a factory with such great exposure of fenestration will rarely require a full lighting load

100% of the time. It is more likely that at the time of the peak sun load, only 20,000, or 15,000 watts will be in use while the footcandle level is maintained by natural illumination. Correct judgement carefully applied to *usage factors* may substantially reduce the load and at the same time, provide a better system.

Stratification

In buildings with high ceilings, conditioned air is seldom delivered from the ceiling height. By delivering the air from a height of 10 ft and allowing the high ceiling air to stratify, less outside air may be required. In the example — see Figure 4.18 — the cubical contents was 280,000 cu ft; and the cfm outside air was obtained by the equation;

$$\frac{280,000 \text{ cu ft}}{60 \text{ min}} = 4666 \text{ cfm per air change}$$

If the air was delivered at 10 ft instead of 14 ft, the cubical contents would be 200 ft x 100 ft x 10 ft = 200,000 cu ft, then;

$$\frac{200,000 \text{ cu ft}}{60 \text{ minutes}} = 3333 \text{ cfm per air change}$$

This represents a 30% reduction in outside air load.

All of these special considerations are open for debate in an examination. The instructions must be read carefully and the information given must be carefully digested *before* the estimate is begun. If the problem is presented in a simple and general manner it should be answered simply and generally. If the problem is expressed in a more detailed and specific manner, it should be answered accordingly. In any case, wherever the candidate feels that he must assume a datum, he should so state — clearly and legibly — on the examination paper.

BIBLIOGRAPHY

ASHRAE Handbook of Fundamentals, 1972, Chapter 19

Carrier Handbook of Air Conditioning System Design, McGraw-Hill, New York, 1965, Part 1

Olivieri, J.B., *How to Design Heating-Cooling Comfort Systems*, Business News Publishing, Birmingham, Mich., 1971

Trane Air Conditioning Manual, The Trane Co., 1965, Chapter 5

PIPEFITTERS FORMULAS AND TABLES

OFFSETS

    When an obstruction occurs in the way of a pipe run, the fitter must "break" around the obstruction to form an *offset*, thereby continuing the pipe run in the same direction but no longer in the same line. Often, the pipefitter can "eyeball" the job. With the help of a rule, straight-edge or string he can achieve a neater and closer fitting than merely eyeball-ing.

    But there are times where the break is critical and the fitter must resort to calculations. Almost every examination will have questions dealing with pipe fitting problems. Figure 4.21 shows a standard 45° offset bend with the usual terminology. Sometimes the *run* is called the "advance" and some books refer to the *travel* as the "longside". Table 4.12 gives the solutions for offset calculations based on the nomenclature used in Figure 4.21.

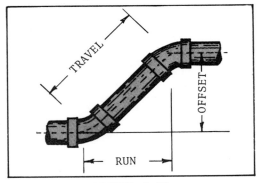

FIGURE 4.21

OFFSET BENDS

    The formulas for several offset bends are given below. Most examinations will have some question on a double offset 540° expansion bend or a 360° expansion U-bend. The calculations of offset bends are always a function of the radius; the radius will usually be given and you will be asked to find the other dimensions. See Table 4.16 for calculated developed lengths for various bends and radii.

250

TABLE 4.12

FORMULAS FOR OFFSET CALCULATIONS

To Find	Multiply	For 45° Ells	For 22-1/2° Ells	For 11-1/2° Ells	For 5-5/8° Ells	For 60° Ells	For 30° Ells	For Angle Bends
Travel	Offset x	1.414	2.613	5.126	10.207	1.155	2.0	Cosec
Offset	Travel x	0.707	0.383	0.195	0.098	0.866	0.5	Sin
Run	Offset x	1.0	2.414	5.027	10.158	0.577	1.732	Cot
Offset	Run x	1.0	0.414	0.199	0.098	1.732	0.577	Tan
Travel	Run x	1.414	1.082	1.019	1.005	2.00	1.155	Sec
Run	Travel x	0.707	0.933	0.981	0.995	0.500	0.866	Cos

1. 240° Crossover bend

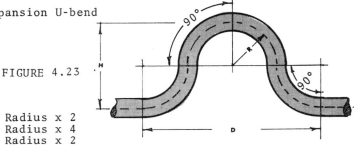

    FIGURE 4.22

    Offset = Radius
    Run   = Radius x 3.464
    Developed length of pipe = Radius x 4.189

2. 360° Expansion U-bend

    FIGURE 4.23

    Spread = Radius x 2
    D   = Radius x 4
    H   = Radius x 2
    Developed length of pipe = Radius x 6.283

3. 540° Double offset expansion bend

    FIGURE 4.24

    Spread = Radius x 0.8284
    D   = Radius x 2.828
    H   = Radius x 3.414
    Developed length of pipe = Radius x 9.425

4. 180° U-bend

FIGURE 4.25

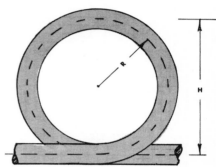

Spread = Radius x 2
H   = Radius
Developed length of pipe = Radius x 3.142

5. 360° Circle bend

FIGURE 4.26

Spread = Radius x 2
H   = Radius x 2
Developed length of pipe = Radius x 6.283

6. 3 Piece 90° turn

    In the layout for fabrication of a 3 piece 90° turn there are two mitered joints. Angles of cuts and pipe lengths are figured as follows:

Angle of cut = $\dfrac{\text{number of degrees of turn}}{2 \times \text{number of welds}}$ or,

$$\frac{90}{2 \times 2} = \frac{90}{4} = 22\text{-}1/2°$$

a = Radius x 0.414
A = 2a = Radius x 0.828
Developed length = 3 A
Cut back distance =
  pipe O.D. x 0.414 =

Example: From Figure 4.27 assuming 4" schedule 40
          pipe and radius of 12 in.;

```
a = 12 in. x 0.414                   =    4.97 in.
A = 2 x 4.97                         =    9.94 in.
Developed length - 3 x 9.94         =   29.82 in.
Cut back = 4.5 x 0.414              =    1.86 in.
B = 9.94 + 1.86                      =   11.80 in.
C = 9.94 - 1.86                      =    8.08 in.
```

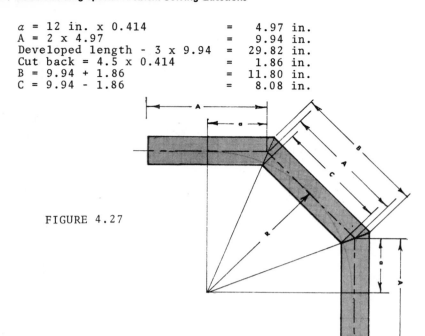

FIGURE 4.27

7.   Three run 45° equal spread

FIGURE 4.28

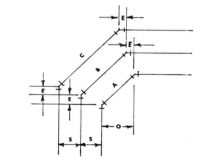

```
A = Offset x 1.414
E = Spread x 0.414
B = A + 2E
C = B + 2E
```

TABLE 4.13

LAYING PROJECTIONS FOR BUTT WELDING ELBOWS
FOR STANDARD AND EXTRA STRONG STEEL PIPE.

This Table may be used to find the
actual pipe length between fittings
by subtracting the projections on
each side from the calculated length.

Nominal Diameter Inches	Dimension for A (inches)		
	Long Radius 90° Elbow	Short Radius 90° Elbow	45° Elbow
1	1-1/2	1	7/8
1-1/4	1-7/8	1-1/4	1
1-1/2	2-1/4	1-1/2	1-1/8
2	3	2	1-3/8
2-1/2	3-3/4	2-1/2	1-3/4
3	4-1/2	3	2
3-1/2	5-1/4	3-1/2	2-1/4
4	6	4	2-1/2
5	7-1/2	5	3-1/8
6	9	6	3-3/4
8	12	8	5
10	15	10	6-1/4

PITCHING PIPE

> *To Find the Drop in a Pipe Run:* multiply the pitch per foot x the length of run -

What is the drop for a 100 ft run of pipe with a 1/4 in. pitch?

100 x 1/4 in. = 100 x .25 = 25 in., 25 ÷ 12 in. = 2.08 ft.

> *To Find the Pitch in a Run of Pipe:* divide the drop (inches) by the length of run (feet).

What is the pitch for a 100 ft run of pipe with a 25 in. drop?

$$\frac{25 \text{ in.}}{100 \text{ ft}} = .25 \text{ inch} = 1/4 \text{ in. pitch}$$

EXPANSION OF PIPE

To calculate the expansion and contraction of pipe it is necessary to know the *coefficient of expansion* for the particular material. Table 4.14 lists various coefficients.

TABLE 4.14

COEFFICIENT OF EXPANSION OF PIPE

METAL	COEFFICIENT OF EXPANSION PER °F
Cast Iron	0.0000059
Copper	0.0000094
Steel	0.0000061
Wrought Iron	0.0000069
Brass	0.0000104

Once the coefficient of expansion is known the formula can be applied;

$$E = C \times L \times \Delta T$$

where

E = Expansion of pipe (inches)
C = Coefficient of expansion
L = Length of run (inches)
ΔT = Temperature change

When the temperature change is not exactly known, it is customary to use a 100°F rise for standard cooling and heating work.

The movement of a 3 inch copper hot-water pipe, 100 ft long when the temperature changes from 70°F to 170°F is equal to;

0.0000094 x 100°F x 100 ft x 12 in. = 1.128 inches

Sizing the Expansion Loop: Any pipe subject to temperature change will expand and contract with each change in a lengthwise direction, in accordance with the above formula. It is therefore essential to provide an expansion section. Table 4.15 is taken from Mueller Brass Co. data; it lists the radii required for different sizes of copper tube to take up to 6 inches of expansion.

TABLE 4.15

| Pipe Size Inches | Radius in Inches (R) | | | | | | | | |
| | For Expansion Of | | | | | | | | |
	1/2"	1"	1-1/2"	2"	2-1/2"	3"	4"	5"	6"
3/4	10	15	19	22	25	27	30	34	38
1	11	16	20	24	27	29	33	38	42
1-1/4	11	17	21	26	29	32	36	42	47
1-1/2	12	18	23	28	31	35	39	46	51
2	14	20	25	31	34	38	44	51	57
2-1/2	16	22	27	32	37	42	47	56	62
3	18	24	30	34	39	45	53	60	67
4	20	28	34	39	44	48	58	66	75
5	22	31	39	44	49	54	62	70	78
6	24	34	42	48	54	59	68	76	83

Bends to the left of the stepped line may be made from 20 ft lengths or less. Bends requiring more than 20 ft lengths must be made up in sections and assembled with couplings or flanges. The expansion section should be cold sprung approximately one-half of the expected distance of expansion

After the number of inches of expansion has been determined from the above formula, the required size expansion loop may be selected from Table 4.15. Still using the above example (the expansion was calculated at 1.128 inches) interpolating from Table 4.15 the required radius for 3 in. pipe is 26 inches. Figures 4.29 and 4.30 show the pipe shape and length for expansion bends and offsets.

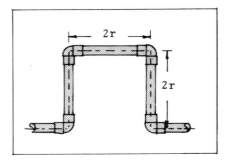

OFFSET WITH FOUR 90° CAST ELLS
FIGURE 4.29

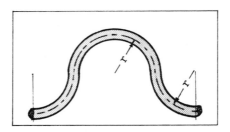

EXPANSION LOOP
FIGURE 4.30

Thus, from Figure 4.29, 13.0 ft of pipe will be required and from Figure 4.30, 13.5 ft of pipe will be required.

If the example called for steel pipe rather than copper the calculations would be somewhat different. Using the Formula $E = C \times L \times \Delta T$;

0.0000061 x 100°F x 100 ft x 12 in. = 0.732 in. expansion

For steel pipe to 20 inches, and expansion of up to 10 inches, the expansion loop may be calculated from Figure 4.31.

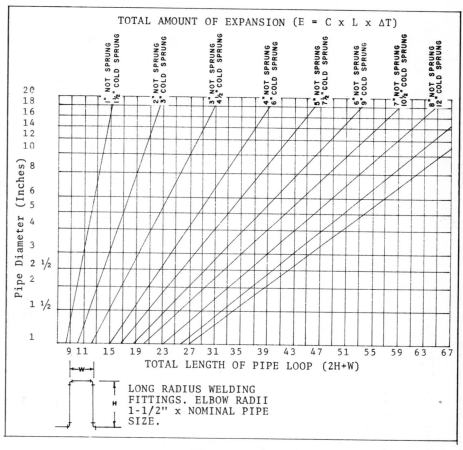

FIGURE 4.31

Given: A 100 ft run of 3 in. steel pipe at 100° F rise.

Find: 1. Expansion in inches
        2. Size of expansion loop

Step 1. From Table 3.4, the coefficient of expansion for steel pipe is 0.0000061
Step 2. Substituting for E = CxLxΔT;
         0.0000061 x 100°F x 100 ft x 12 in. = 0.732 in. expansion
Step 3. Enter the left ordinate of the chart in Figure 3.12 at 3 in. pipe diameter. Follow the horizontal line to the curve for 1" not sprung. Extrapolating the curve, the required length of the loop is 10 ft. Maintaining the recommended ratio;

$$W = \frac{H}{2} = \frac{4}{2} \text{, therefore, } W=2', H=4', H=4'.$$

TABLE 4.16

CALCULATED LENGTHS OF PIPE FOR VARIOUS BENDS

RADIUS OF PIPE BENDS		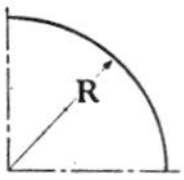	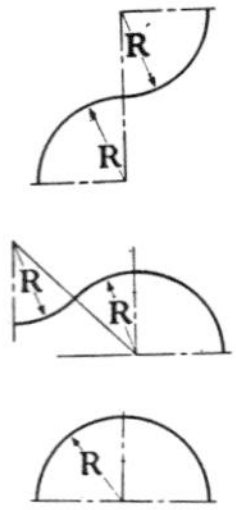	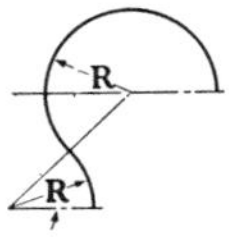	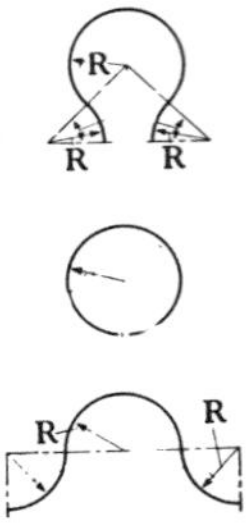	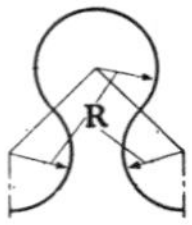
INCHES	FEET	90°	180°	270°	360°	540°
1		1½"	3"	4¾"	6¼"	9½"
2		3	6¼	9½	12½	18¾
3	¼	4¾	9½	14¼	18¾	28¼
4		6¼	12½	18¾	25¼	37¾
5		7¾	15¾	23½	31½	47¼
6	½	9½	18¾	28¼	37¾	56½
7		11	22	33	44	66
8		12½	25¼	37¾	50¼	75½
9	¾	14¼	28¼	42½	56½	84¾
10		15¾	31½	47¼	62¾	94¼
11		17¼	34½	51¾	69	103¾
12	1	18¾	37¾	56½	75½	113
12.5		19¾	39¼	59	78½	117¾
14		22	44	66	88	132
15	1¼	23½	47	70¾	94¼	141¼
16		25¼	50¼	75½	100½	150¾
17.5		27½	55	82½	110	165
18	1½	28¼	56½	84¾	113	169¾
20		31½	62¾	94¼	125¾	188½
21	1¾	33	66	99	132	198
24	2	37¾	75½	113	150¾	226¼
25		39¼	78½	117¾	157	235½
30	2½	47¼	94¼	141¼	188½	282¾
32		50¼	100½	150¾	201	301½
36	3	56½	113	169½	226¼	339¼
40		62¾	125¾	188½	251¼	377
48	4	75½	150¾	226¼	301½	452½
50		78½	157	235½	314¼	471¼
56		88	176	264	351¾	527¾
60	5	94¼	188½	282¾	377	565½
64		100½	201	301½	402	603¼
70		110	220	329¾	439¾	659¾
72	6	113	226¼	339¼	452½	678½
80		125¾	251¼	377	502¾	754
84	7	132	263¾	395¾	527¾	791½
90	7½	141¼	282¾	424	565½	848¼
96	8	150¾	301½	452½	603	904¾
100		157	314¼	471¼	628¼	942½
108	9	169½	339¼	509	678½	1017¾
120	10	188½	377	565½	754	1131
132	11	207¼	414¾	622	829½	1244
144	12	226¼	452½	678½	904¾	1357¼
156	13	245	490	735¼	980¼	1470¼
168	14	263¾	527¾	791½	1055½	1583½
180	15	282¾	565½	848¼	1131	1696½
192	16	301½	603	904¾	1206¼	1809½
204	17	320½	640¾	961¼	1281¾	1922½
216	18	339¼	678½	1017¾	1357¼	2035¾
228	19	358	716¼	1074½	1432½	2148¾
240	20	377	754	1131	1508	2262

BIBLIOGRAPHY

Bachman, G.K., *Pipefitter's and Plumber's Vest Pocket Reference Book*, Prentice-Hall, New Jersey, 1960

Gladstone, J., *Journeyman General Mechanical Exam.*, Engineer's Press, Coral Gables, Fla., 1973, Chapter 7

Lindsey, F.R., *Pipefitters Handbook*, 3rd ed., Industrial Press, New York, 1967

AIR BALANCING AND SYSTEM TESTING

The concept of professional testing and balancing is recent to the trade, and questions dealing with this subject have only recently been appearing on examination. Consequently, the frame of reference is very shallow, and it is difficult to gather a good sampling of representative questions.

During the last two years training courses in *Test and Balance* have been organized by SMACNA, MCAA and various apprenticeship programs, and in some cases elaborate test labs have been set up for educational purposes. But no general pattern for examination questions has yet emerged, and it will probably be another couple of years before such information can be syncretized.

Most examinations for air conditioning master contractor are written by persons who have only cursory knowledge of this specialized function and thus far, questions are usually limited to fan laws. Sometimes, however, questions may be more complicated; the questions shown later in this chapter were taken from an actual examination in Dade County.

The skilled balancing technician must understand fan operating characteristics, the theory of fluid flow, fan and pulley laws, how to use the variety of instruments required to take flow and pressure readings, how to adjust speed, flow and pressure, etc. System start-up, test and balance, and proper maintenance, require a thorough understanding of the mechanics and laws of pulleys and belts; and although fan law solutions can be worked out with great theoretical accuracy on paper, such fan laws are worthless unless a skilled technician can translate them into actual working conditions in the field.

BELTS, PULLEYS AND PULLEY LAWS

Drive sets for fans and blowers consist of a driver pulley on the motor shaft, a driven pulley on the blower shaft, and a belt or set of matched belts to transmit the power. Pulley formulas are usually given in pulley diameters; for accuracy they should be considered in actual pitch diameters.

The four basic pulley laws are:

1. Rpm of driven = $\dfrac{\text{diameter of driver x rpm}}{\text{diameter of driven}}$

2. Rpm of driver = $\dfrac{\text{diameter of driven x rpm}}{\text{diameter of driver}}$

3. Diameter of driven = $\dfrac{\text{diameter of driver x rpm}}{\text{rpm of driven}}$

4. Diameter of driver = $\dfrac{\text{diameter of driven x rpm}}{\text{rpm of driver}}$

Example: The diameter of a motor pulley is 6 inches, the diameter of the fan pulley is 12 inches, and the speed of the motor is 3800 rpm. What is the speed of the fan?

$$\frac{6 \text{ inches x } 3800 \text{ rpm}}{12 \text{ inches}} = 1900 \text{ rpm}$$

V-Belts

All standard V-belts are identified by a standard numbering system. *Single V-belts* consist of a letter numeral combination designating length of belt. An 8.0 in. belt is designated 2L080, a 52.0 in. belt is designated 3L520, and a 100.0 in. belt is designated 4L100. The first digit indicates the number of digits in the inch length. *Variable V-Belts* consist of a standard numbering system that indicates the nominal belt width in sixteenths of an inch by the first two numbers. The third and fourth numbers indicate the angle of the groove in which the belt is designed to operate. The digits following the letter "V" indicate the pitch length to the nearest 1/10 in. A belt numbered 1430V450 is a V-belt of 14/16 in. nominal top width designed to operate in a sheave of 30 degrees (groove angle) and have a pitch length of

45.0 in. Length tolerances for belts under 100 in. should be $\pm$ 0.0025; for over 100 in., $\pm$ 0.0075.

The length of a V-belt is calculated by the formula:

$$L = 2C + 1.57\ (D+d) + \frac{D-d^2}{4C}$$

where C = Center distance between shafts, inches
D = Outside diameter of large sheave, inches
d = Outside diameter of small sheave, inches

The center distance between two shafts is given by the formula:

$$C = \frac{K + \sqrt{K^2 - 32\ (D-d)^2}}{16}$$

where $K = 4L - 6.28\ (D+d)$
L = Length of belt, inches
D = Outside diameter of large sheave, inches
d = Outside diameter of small sheave, inches

The belt speed for various rpm's may be found where the sheave pitch diameter is known, from the formula:

$$fpm = pd \times rpm \times 0.262$$

MOTORS, FANS AND FAN LAWS

For the test and balance technician, there are 5 basic fan laws that are absolutely essential to his work. With these formulas, the technician can adjust speeds to meet any new set of conditions.

I. Cfm varies in direct proportion to rpm.

$$cfm_2 = cfm_1 \times \frac{rpm_2}{rpm_1}$$

II. Sp varies as the square of the rpm.

$$Sp_2 = Sp_1 \times \left(\frac{rpm_2}{rpm_1}\right)^2$$

III. HP varies as the cube of the rpm.

$$hp_2 = hp_1 \times \left(\frac{rpm_2}{rpm_1}\right)^3$$

Example: A fan is operating at 0.5 in. static pressure at a speed of 1000 rpm and moving 2000 cfm of air at 1.5 hp. Using the above 3 formulas, find the new Sp, rpm, and hp if the design calls for 2200 cfm.

1. $rpm_2 = rpm_1 \times \dfrac{cfm_2}{cfm_1} = 1000 \times \dfrac{2200}{2000} = 1100 \ rpm_2$

2. $Sp_2 = Sp_1 \times \left(\dfrac{cfm_2}{cfm_1}\right)^2 = 0.5 \times \left(\dfrac{2200}{2000}\right)^2 = 0.6 \ Sp_2$

3. $hp_2 = hp_1 \times \left(\dfrac{cfm_2}{cfm_1}\right)^3 = 1.5 \times \left(\dfrac{2200}{2000}\right)^3 = 2 \ hp_2$

IV. Amperage varies as the cube of the cfm.

$$amps_2 = amps_1 \times \left(\dfrac{cfm_2}{cfm_1}\right)^3$$

Example: A fan is handling 2000 cfm and drawing 2.5 amps. What will be the amperage reading necessary to bring the fan up to 2200 cfm?

$$amps_2 = amps_1 \times \left(\dfrac{cfm_2}{cfm_1}\right)^3 = 2.5 \times \left(\dfrac{2200}{2000}\right)^3 = 3.33 \ amps_2$$

V. Horsepower varies as the square root of the pressure ratio cubed.

$$hp_2 = hp_1 \times \sqrt{\left(\dfrac{Sp_2}{Sp_1}\right)^3}$$

Example: A fan is operating at 0.5 in. static pressure and 1.5 horsepower. What will the new horsepower be if the static pressure is increased to 0.6 in.?

$$hp_2 = hp_1 \; x \; \sqrt{\left(\frac{Sp_2}{Sp_1}\right)^3} = 1.5 \; x \; \sqrt{\left(\frac{.6}{.5}\right)^3} = 2 \; hp_2$$

Formulas for Adjusting Sheaves

1. Given a change in cfm, find the new pulley setting.

$$pd_2 = pd_1 \; x \; \frac{cfm_2}{cfm_1}$$

2. Given a maximum brake horsepower, find the new pitch diameter required to change from an existing pitch diameter.

$$pd_2 = pd_1 \; x \; \sqrt[3]{\frac{max \; bhp_2}{bhp_1}}$$

where pd = pitch diameter
       bhp = brake horsepower
       cfm = air quantity at the fan.

TABLE 4.17

CONVERSION OF VELOCITY PRESSURES (Vp) TO VELOCITY (fpm).

FROM THE FORMULA: FPM = $4005\sqrt{Vp}$

VELOCITY PRESSURE IN. OF H_2O	VELOCITY FPM	VELOCITY PRESSURE IN. OF H_2O	VELOCITY FPM	VELOCITY PRESSURE IN. OF H_2O	VELOCITY FPM
0.0056	300	0.100	1300	0.681	3300
0.0060	310	0.105	1350	0.695	3350
0.0064	320	0.122	1400	0.722	3400
0.0068	330	0.131	1450	0.740	3450
0.0072	340	0.140	1500	0.766	3500
0.0076	350	0.149	1550	0.785	3550
0.0081	360	0.160	1600	0.810	3600
0.0086	370	0.169	1650	0.825	3650
0.0090	380	0.181	1700	0.860	3700
0.0095	390	0.190	1750	0.875	3750
0.010	400	0.208	1800	0.90	3800
0.011	425	0.213	1850	0.92	3850
0.012	450	0.225	1900	0.95	3900
0.014	475	0.237	1950	0.97	3950
0.016	500	0.250	2000	1.00	4000
0.017	525	0.262	2050	1.02	4050
0.019	550	0.276	2100	1.04	4100
0.021	575	0.290	2150	1.06	4150
0.022	600	0.302	2200	1.08	4200
0.024	625	0.319	2250	1.12	4250
0.026	650	0.331	2300	1.14	4300
0.028	675	0.348	2350	1.17	4350
0.030	700	0.360	2400	1.20	4400
0.033	725	0.375	2450	1.23	4450
0.035	750	0.391	2500	1.25	4500
0.038	775	0.408	2550	1.28	4550
0.040	800	0.422	2600	1.31	4600
0.042	825	0.439	2650	1.34	4650
0.045	850	0.456	2700	1.37	4700
0.048	875	0.467	2750	1.40	4750
0.050	900	0.490	2800	1.43	4800
0.053	925	0.509	2850	1.46	4850
0.057	950	0.530	2900	1.50	4900
0.059	975	0.543	2950	1.53	4950
0.062	1000	0.562	3000	1.56	5000
0.066	1050	0.576	3050	1.59	5050
0.075	1100	0.600	3100	1.62	5100
0.082	1150	0.615	3150	1.65	5150
0.090	1200	0.640	3200	1.68	5200
0.097	1250	0.665	3250	1.72	5250

FIGURE 4.32

SPEED-O-GRAPH

CONVERSION: VELOCITY, STATIC PRESSURE, VELOCITY PRESSURE,
TOTAL PRESSURE

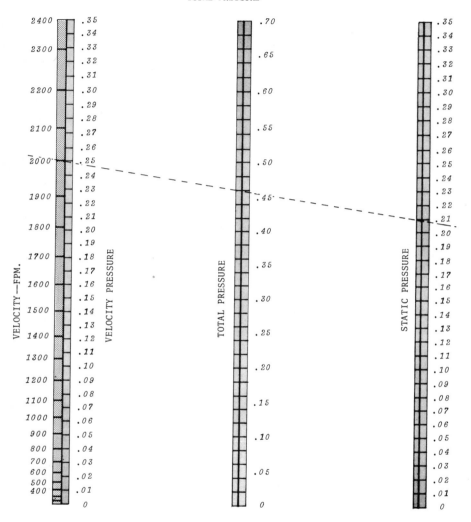

Given: Duct velocity, 2000 fpm; static pressure, .21 in.
H₂0 *Find:* Velocity pressure and total pressure.
1. Lay a straight edge across the two outer scales inter-
secting .21 on the Sp scale and 2000 on the velocity scale.
2. Read the answer = .25 Vp and .46 Tp. Note: When
laying the straight edge, cross the scales on the heavy
center vertical line.

PRESSURE EQUIVALENTS

Fan static pressures are usually given in terms of *inches of water* rather than *psi* because a small decimal such as 0.036 psi is more cumbersome to work with than its equivalent, 1.0 inch of water. Pressure and head are **equivalent** terms; 1 psi corresponds to 27.7 inches of water.

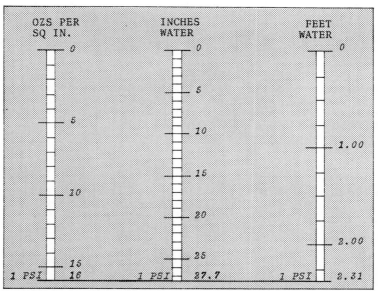

Oz/in.² = ounces per square inch   In. H₂O = inches of water @ 68F
In.Hg = inches of mercury @ 0C   Psi = pounds per square inch
Ft H₂O = feet of water @ 68F   Psf = pounds per square foot
kg/cm² = kilograms per square **centimeter**

MULTIPLY	X	TO OBTAIN
Psi	16	Oz/in.²
Psi	2.31	Ft H₂O
Psi	27.73	In. H₂O
Psi	0.0703	Kg/cm ²
Psi	2.036	In. Hg
In. H₂O	0.07342	In. Hg
In. H₂O	0.5770	Oz/In.²
In. H₂O	0.03606	Psi
In. H₂O	5.196	Psf
Ft H₂O	0.4328	Psi
Ft H₂O	62.32	Psf
TO OBTAIN ABOVE	DIVIDE ABOVE	STARTING WITH ABOVE

FIGURE 4.33

SAMPLE TEST QUESTIONS FOR TEST AND BALANCE

The following questions appeared on an examination for Master of Air Conditioning in Dade County, the answers are given at the end of the questions. All questions are based on 29.92" barometric pressure.

1. A Pitot tube traverse of a 4 sq ft duct gives a reading of 2" Tp and 1.50" Sp. The cfm flowing through the duct is _____.

   (a) 12,200   (c) 10,960.2

   (b) 11,400   (d) 8,000

2. A Pitot tube traverse reads 0.213" Vp on an inclined draft gage and the air flow measures 115 F on a thermometer. At an altitude of 3500 ft above sea level, the actual velocity in fpm is _____.

   (a) 1850   (c) 1498

   (b) 1670   (d) 1262

3. When making a standard 16 point duct traverse for an 18"x18" square duct, the first hole drilled should be _____ inches away from the duct wall.

   (a) 4.50   (c) 9.75

   (b) 18.0   (d) 2.25

4. A system is using 7.5 hp and handling 10,000 cfm at 650 rpm and 1.5" Sp, if the motor is changed to 15 hp the cfm will be _____.

   (a) 12,600   (c) 10,260

   (b) 10,620   (d) 16,200

5. Which of the following instruments would *not* be used in conjunction with a Pitot tube?

   (a) Magnahelic gage   (c) Anemometer

   (b) Manometer   (d) Draft gage

6. If you had a completely equipped combustion kit which of the following text functions could you *not* perform?

    (a) CO_2 content of flue gas

    (b) Excess combustion air

    (c) Discoloration of refrigerated meat

    (d) Fuel oil pressure at pump

7. A Pitot tube traverse of a duct with a square elbow shows a Tp of .515" H_2O and a velocity pressure of 2.50" H_2O, therefore, the Sp will be _____.

    (a) 1.985   (c) 3.015

    (b) Minus 1.985   (d) Minus 3.015

8. A ceiling diffuser has a "K" factor of .80 and a neck area of 2 sq ft, if the diffuser is known to handle 500 cfm, the jet velocity will be _____ fpm.

    (a) 1000   (c) 800

    (b) 225   (d) 625

9. In test and balance work the following eight figures are used daily by the technician. Match the numbers on the right to correspond with the correct letters on the left.

    (a) 8.33   1. Standard air density
    (b) 5.196   2. 60 Min x 8.33 lb/gal lb of 1 fpm
    (c) 12,000   3. Wt of 1 gpm standard H_2O
    (d) 15,000   4. Ft of H_2O @ 1 lb pressure
    (e) 3421   5. Lb/sq ft/in. of H_2O pressure
    (f) 2.31   6. Btu/hr/ton refrigeration
    (g) .075   7. Btu/hr/kw
    (h) 500   8. Btu/hr/cooling tower standard ton

10. In the figure below, four Pitot tube connections are
     shown. Which is correct?

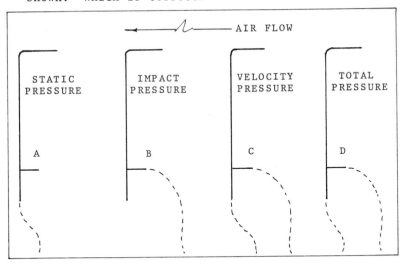

ANSWERS TO QUESTIONS ON TEST AND BALANCE

1. B   Tp = Sp+Vp, so Vp = 2.0" minus 1.50" = .50"
           From Table 4.17 .50" = 2850 fpm
           2850 fpm x 4 = 11,400 cfm

2. C   From Table 4.17 .213" Vp = 1850 fpm
           From Figure 4.15 115°F = .92
           From Figure 4.14 3500 ft = .88
           Fpm standard air = 1850 x .92 x .88 = 1498 fpm

3. D   2.25"

4. A $\dfrac{HP_1}{HP_2} = \left(\dfrac{CFM_1}{CFM_2}\right)^3 = \sqrt[3]{\dfrac{HP_2}{HP_1} \text{ x } CFM_1} =$

           1.26 x 10,000 = 12,600 CFM₂

5. C   Anemometer

6. D Fuel oil pressure at pump

7. B Tp = Sp+Vp $\therefore$ Sp = Tp $-$ Vp = .515 $-$ 2.50 =

   minus 1.985 Sp

8. D "K" x fpm = cfm $\therefore$ cfm $\div$ "K" = fpm, so

   500 $\div$ .8 = 625 fpm. The neck area is irrelevant

9. A = 3   E = 7
   B = 5   F = 4
   C = 6   G = 1
   D = 8   H = 2

10. C Velocity pressure

BIBLIOGRAPHY

Air Velocity Instruments, Bulletin H-100, Dwyer Instrument Co.

ASHRAE Handbook of Fundamentals, 1972, Chapter 12.

ASHRAE Handbook and Product Directory: Systems, 1973, Chapter 40.

Balancing Manual, BM 9-70, Tuttle and Bailey.

Fan Engineering, Buffalo Forge, 1970.

Gladstone, J., *Air Conditioning Testing and Balancing: A Field Practice Manual*, Van Nostrand Reinhold, N.Y. 1974.

Kahoe, H.T., "How to Test and Balance Air Conditioning Systems", *Air Conditioning and Refrigeration Business*, Nov. 1967, p.45.

Manual for the Balancing and Adjustment of Air Distribution Systems, SMACNA, 1967.

Trane Air Conditioning Manual, The Trane Co., 1965, Chapter 9.

Uni-Flow Air Distribution Balancing Bulletin, Barber Coleman.

Wind, M. "Air Balancing and Testing," *Heating, Piping and Air Conditioning*, Dec. 1968, p.112.

5. GENERAL SAFETY REQUIREMENTS

Safety regulations may vary slightly in different areas but the major basic rules are universal. The material in this section is derived mainly from three sources; *Safety Manual*, Corps of Engineers U.S. Army, *Safety Regulations*, Florida State Law 8AS, and *Manual of Crane Operation and Hitching*, Allis-Chalmers Manufacturing Company. It covers most of the material encountered in mechanical contractors' examinations as well as journeymen's.

Care and Use of Portable Ladders.-

(a) In General. To get maximum serviceability, safety, and to eliminate unnecessary damage of equipment, good safe practices in the use and care of ladder equipment must be employed by the users.

(b) Care of Ladders

(1) Ladders shall be maintained in good condition at all times, the joint between the steps and side rails shall be tight, all hardware and fittings securely attached, and the movable parts shall operate freely without binding or undue play.

(2) Metal bearings of locks, wheels, pulleys, etc., shall be frequently lubricated.

(3) Frayed or badly worn rope shall be replaced.

(4) Safety feet and other auxiliary equipment shall be kept in good condition to insure proper performance.

(5) Ladders should be stored in such a manner as to provide ease of access or inspection, and to prevent danger of accident when withdrawing a ladder for use.

(6) Wood ladders, when not in use, should be stored at a location where they will not be exposed to the elements, but where there is good ventilation. They shall not be stored near radiators, stoves, steam pipes, or other places subjected to excessive heat or dampness.

(7) Ladders stored in a horizontal position should be supported at a sufficient number of points to avoid sagging and permanent set.

(8) Ladders carried on vehicles should be adequately supported to avoid sagging and securely fastened in position to minimize chafing and the effects of road shocks. Tying the ladder to each support point will greatly reduce damage due to road shock.

(9) Wood ladders should be kept coated with a transparent protective material. Wood ladders shall not be painted with an opaque pigmental material.

(10) Ladders shall be inspected frequently and those which have developed defects shall be withdrawn from service for repair or destruction.

(11) Rungs shall be kept free of grease and oil.

(c)   Use of Ladders

(1)   Portable rung and cleat ladders shall, where possible, be used at such a pitch that the horizontal distance from the top support to the foot of the ladder is one-quarter of the working length of the ladder (the length along the ladder between the foot and the top support). The ladder shall be so placed as to prevent slipping, or it shall be lashed, or held in position. Ladders shall not be used in a horizontal position as platforms, runways, scaffolds.

(2)   Crowding on ladders shall not be permitted. Portable ladders are designed as a one-man working ladder based on a 200-pound load.

(3)   Portable ladders shall be so placed that the side rails have a secure footing. Safety shoes of good substantial design should be installed on all ladders. Where ladders with no safety shoes or spikes are used on hard, slick surfaces, a foot ladder board should be employed. The top of the ladder must be placed with the two rails supported, unless equipped with a single support attachment. Such an attachment should be substantial, and large enough to support the ladder under load.

(4)   Ladders shall not be placed in front of doors opening toward the ladder unless the door is blocked open, locked, or guarded.

(5)   Ladders shall not be placed on boxes, barrels, or other unstable bases to obtain additional height.

(6)   To support the top of a ladder, at a window opening, a board should be lashed across the back of the ladder, extending across the window and providing firm support against the building walls or window frames.

(7)   When ascending or descending, the user shall face the ladder.

(8)   Ladders with broken or missing steps, rungs, or cleats, broken side rails, or other faulty equipment shall not be used. Improvised repairs shall not be made.

(9)   Short ladders shall not be spliced together to provide long sections.

(10)   Ladders made by fastening cleats across a single rail shall not be used.

(11)   In building construction, where warranted by height or operations or traffic conditions, separate ladders shall be designated for ascent and descent.

(12)   Improper Use. Ladders should not be used as a brace, skid, guy or gin pole, gangway, or for other uses than that for which they were intended, unless specifically recommended for use by the manufacturer.

(13)   Tops of the ordinary types of step ladders shall not be used as teps

(14)   On two-section extension ladders the minimum overlap for the two sections in use shall be as follows:

Size of Ladder (Feet)	Overlap (Feet)
Up to and including 36	3
Over 36 up to and including 44	4

(15) The back section of a step ladder is for the support of the front section and shall not be used for ascending or descending.

(16) Where it is necessary to install a gang ladder wide enough to permit traffic in both directions at the same time, a center rail shall be provided. One side of the ladder should be plainly marked "up" and the other side "down".

(17) Electrical hazards. All metal ladders are electrical conductors. They shall not be used in the vicinity of electrical equipment.

(18) The side rails of all portable ladders shall extend not less than three and one-half (3-1/2) feet above the platform or floor served. The ladder should be so placed that the landing rung is at or slightly above the floor or platform.

(19) Areas where portable ladders are used shall be kept clear of rubbish and waste materials. Unused materials shall be safely stored.

Boatswains' Chairs.-

(a) The chair shall be a seat not less than two feet long by one foot wide and one inch thick.

(b) Cleats shall be nailed to the underside of each end of the chair and shall project at least nine inches in front of the seat.

(c) The chair shall be supported by means of a sling attached to a suspension rope. The entire assembly shall have a safety factor of four.

(d) The suspension rope shall either be securely fastened to a fixed object overhead, or passed through an overhead block securely fastened. The free end shall be securely fastened to a fixed and easily accessible object, and the chair raised or lowered, if necessary with the aid of helpers.

(e) When the suspension rope is attached by means of a hitch, the workman shall be provided with stirrups upon which he can rest his weight while he is shifting the hitch by which the chair is made fast, and the stirrups shall be supported independently of the chair itself.

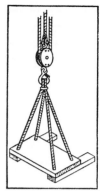

FIGURE 5.1

(f)   Every workman using a boatswains' chair shall be provided with a safety belt secured to the supporting tackle.

(g)   When a boatswains' chair is used by a workman using a torch, or any open flame, fiber rope slings shall not be used. The slings shall be at least three-eighths inch wire rope.

Tie Beam Jack Scaffold.-

(a)   This type of scaffold shall be for light work inside buildings and shall be composed of framed wood jacks supporting a plank platform or construction of equivalent strength.

(b)   The maximum height of the jacks shall be 60 inches, the base shall be not less than 30 inches and the top member not less than 20 inches long.

(c)   If the jacks are framed up of lumber, the top member, the inside upright (the one against the wall) and the outside sloped member shall be of 2 x 4 inch lumber and the bottom member of 1 x 6 inch boards.

(d)   At the top of the jack, in addition to the 2 x 4 member, two 1 x 6 inch boards shall be nailed parallel to the top member and on its sides as bearers.

(e)   A 1 x 6 inch diagonal brace shall extend from near the bottom of the upright member to near the top of the outside sloped member.

(f)   Platform planks shall be nailed to the top member of each jack.

(g)   A 1 x 4 inch logitudinal brace shall be nailed along the outside of each jack at floor level.

(h)   Jacks shall be secured to the walls of the building to prevent toppling and wedged against end walls to prevent lateral displacement.

(i)   The jacks shall be set up not more than eight feet apart.

(j)   When this type of scaffold is used, it shall not be built up more than one jack high and the scaffold shall be set up on a level and stable foundation.

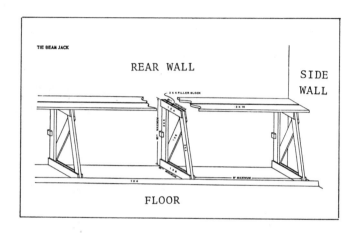

TIE BEAM JACK

REAR WALL

SIDE WALL

FLOOR

FIGURE 5.2

Hand Signal for Crane Operation

Figure 5.3 gives the standard hand signals for crane operation as recommended by The American Society of Mechanical Engineers, *Standard B30.5-1968, Crawler, Locomotive and Truck Cranes.* Reprinted by permission of the publisher, The American Society of Mechanical Engineers, 345 East 47th Street, New York.

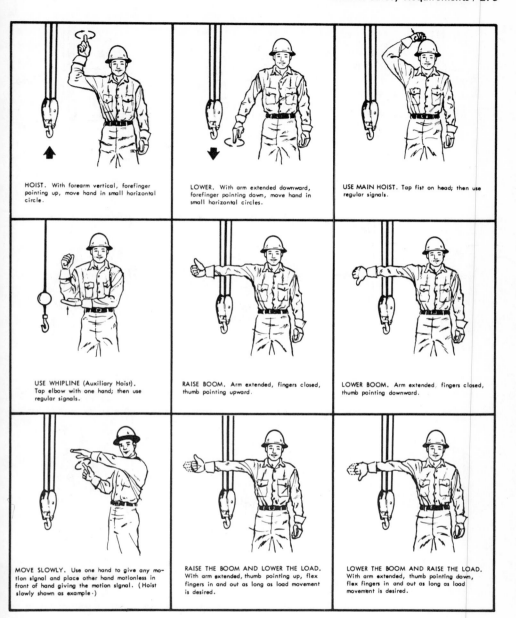

STANDARD HAND SIGNALS FOR CONTROLLING CRANE OPERATIONS

FIGURE 5.3

Reprinted by permission of the publisher, the American
Society of Mechanical Engineer, 345 East 47 St. New York

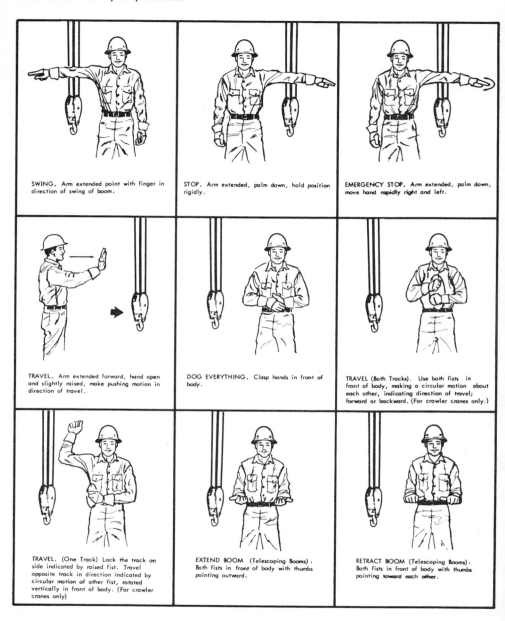

STANDARD HAND SIGNALS FOR CONTROLLING CRANE OPERATIONS

FIGURE 5.3 (continued)

Hitching Equipment and Safe Practice

When using chokers do not pass the running line through an eye splice - use a shackle, and keep the shackle pin in the eye.

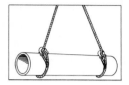

Use shackles for wire rope choker hitches.

FIGURE 5.4

When forming an eye with clamps always have the nuts on the standing part and use a thimble as shown in Figure 5.5.

Use of thimble in eye splice.

FIGURE 5.5

Wire rope slings should be protected from sharp corners with corner irons or heavy chafing gear, Figure 5.6.

Each leg of a wire rope sling should be secured at the hook to prevent reeving of the sling on the hook. The proper method of rigging wire rope slings to the hook is shown in Figure 5.7.

Wire rope should never be knotted and the wire should be visually checked each time before using.

FIGURE 5.7   FIGURE 5.8

Manila rope should be protected against weather, excessive heat, solvents, grease etc., and should be visually inspected each time before using. Manila slings should be protected with chafing gear or bagging as shown in Figure 5.8. A blackwall hitch is shown in Figure 5.9, this is the proper method of rigging a manila rope sling to the hook. The proper steps in making a wrap-around hitch are illustrated in Figure 5.10.

Blackwall
Hitch

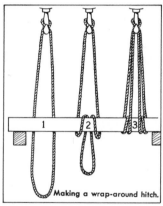

Making a wrap-around hitch.

FIGURE 5.9   FIGURE 5.10

Safe Load Tables

Safe loading for manila and wire rope slings is a function of the number of sling legs connecting the crane hook, and the load angle; the smaller the load angle, the lower the lifting efficiency. Table 5.1 gives the angle between the horizontal surface of the load and the sling as a percentage of a straight lift.

TABLE 5.1

SLING ANGLE EFFICIENCY

LOAD ANGLE Degrees	LIFTING EFFICIENCY Percent
90	100
60	86.6
45	70.7
30	50
15	25
0	0

To use the following Safe Load Tables, see Figure 5.11.
Assuming a packaged chiller weighs 32,000 lbs and the load
angle will be 30°; from Table 5.5 use a 1-1/2 in. diameter
wire rope sling with four legs. Now, dividing the load by
four shackles gives four tons per shackle, but Table 5.1
shows that the lifting efficiency for a 30° angle is 50%.
Therefore, *double* the listed 90° safe load in Table 5.6
and select 1-1/4 in. shackles.

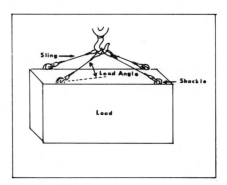

FIGURE 5.11

TABLE 5.2

MANILA ROPE SLINGS - STRAIGHT LEG

TONS

ROPE DIAMETER INCHES	Straight 2 Leg	Straight 4 Leg	Choker 2 Leg	60° Choker 4 Leg	45° Choker 4 Leg	30° Choker 4 Leg
1/2	1/2	1	1/3	2/3	1/2	1/3
3/4	3/4	1-1/2	3/4	1-1/4	1	3/4
1	1-1/2	3	1-1/4	2	1-1/2	1-1/4
1-1/2	3	6	2	4	3	2
2	5	10	4	7	6	4
2-1/2	7	15	6	10	8	6
3	10	20	8	14	12	8
3-1/2	14	29	11	20	16	11
4	17	34	13	23	19	13

TABLE 5.3

MANILA ROPE SLINGS — BASKET

ROPE DIAMETER INCHES	60° Basket 4 Leg	45° Basket 4 Leg	30° Basket 4 Leg	60° Basket 6 Leg	45° Basket 6 Leg	30° Basket 6 Leg
			TONS			
1/2	3/4	2/3	1/2	1	3/4	2/3
3/4	1-1/2	1	3/4	2-1/4	2	1
1	2-1/2	2	1-1/2	4	3	2
1-1/2	5	4	3	8	6	4
2	9	7	5	13	11	7
2-1/2	13	11	7	19	16	11
3	18	15	10	27	22	15
3-1/2	25	21	15	38	31	22
4	30	24	17	44	36	25

TABLE 5.4

WIRE ROPE SLINGS — STRAIGHT LEG

ROPE DIAMETER INCHES	TONS				
	Straight 1 Leg	Choker 1 Leg	60° Choker 2 Leg	45° Choker 2 Leg	30° Choker 2 Leg
1/4	1/2	1/3	2/3	1/2	1/3
3/8	1	3/4	1-1/4	1	3/4
1/2	2	1-1/2	2-1/2	2	1-1/2
5/8	3	2	4	3	2
3/4	4	3	5	4	3
1	7	5	8	7	5
1-1/4	10	7	12	9	7
1-1/2	13	9	16	13	9
2	21	15	27	22	15
2-1/2	28	22	38	31	22
3	36	28	49	40	28
3-1/2	40	34	59	48	34

TABLE 5.5

WIRE ROPE SLINGS — BASKET

ROPE DIAMETER INCHES	60° Basket 2 Leg	45° Basket 2 Leg	30° Basket 2 Leg	60° Basket 4 Leg	45° Basket 4 Leg	30° Basket 4 Leg
			TONS			
1/4	2/3	1/2	1/3	1	1	3/4
3/8	1-1/2	1	3/4	3	2	1-1/2
1/2	2-1/2	2	1-1/2	5	4	3
5/8	4	3	2	7	6	4
3/4	5	4	3	11	9	6
1	9	7	5	18	15	10
1-1/4	13	11	7	26	21	15
1-1/2	17	14	10	35	28	20
2	27	22	15	53	44	31
2-1/2	38	31	22	75	61	43
3	49	40	29	97	80	56
3-1/2	59	49	34	118	97	68

TABLE 5.6
SAFELOAD FOR SHACKLES

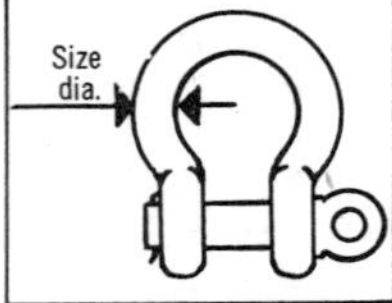

Size	Safe Load at 90°
Inches	Tons
1/4	1/3
5/16	1/2
3/8	3/4
7/16	1
1/2	1 1/2
5/8	2
3/4	3
7/8	4
1	5 1/2
1 1/8	6 1/2
1 1/4	8
1 3/8	10
1 1/2	12
1 3/4	16
2	21
2 1/4	27
2 1/2	34
2 3/4	40
3	50

*Note: When attachments (Shackles, "S" Hooks, etc.) are used at angles other than 90°, reduce safe load rating according to Angle Efficiency Table,

Knots

The origin of most knots, hitches and bends, are lost in the dim antiquity of our seafaring ancestors. Aside from exhibits of the art of seamanship and Boy Scout jamborees, most knots and fancy rope work are now merely of academic interest. Most manuals illustrate dozens of different knots and ties, while in fact, the pipefitter — or rigger — using modern equipment, will seldom need to use more than four knots.

The *bowline* is the king of knots. If made correctly it can *never* jamb. Useful for lowering buckets, tools or men over the side or whenever an absolutely safe hitch is needed. Figure 5.12.

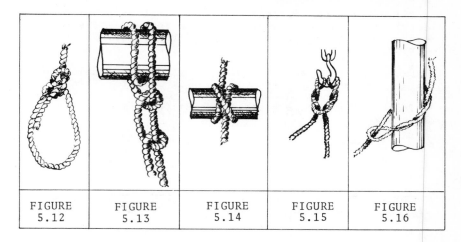

FIGURE 5.12	FIGURE 5.13	FIGURE 5.14	FIGURE 5.15	FIGURE 5.16

The *half hitch, two half hitches,* and the *round turn with two half hitches,* are very useful knots to keep the ends of pipe or timber from spinning, to hold upright objects in line with the pull, and to make fast lines of moderate size. Figure 5.13.

The *clove hitch* is really two half hitches around a stancion or pipe. A very useful knot, it will take great strain and is easy to unbend. Figure 5.14.

The *cat's paw* can be made anywhere on the bight of a rope when it is necessary to clap on a handy billy, or tackle, to the hauling part of a larger purchase. Figure 5.15.

The *timber hitch,* used together with a *half hitch* is good for towing a spar, and hoisting pipe or timber. Figure 5.16.

Grounding of Portable Electrical Equipment

All portable electric tools are to be grounded prior to use. The preferable grounding device shall be so arranged that the ground connection will be established automatically when the portable cable is connected to the fixed wiring. The grounding conductor shall be contained in or on the cable or cord assembly. The grounding conductor shall be contained in or on the cable or cord assembly. The grounding conductor shall be connected to ground, but shall not be a grounded circuit conductor.

All exposed non-current-carrying metal parts of portable electrical equipment operated at more than 50 volts to ground shall be effectively grounded regardless of use or location.

Exception - Electric tools provided with double insulation are exempted.

Table 5.7 offers a guide to the size of conductors for portable tools.

TABLE 5.7

CONDUCTORS FOR PORTABLE POWER TOOLS

HP	50'	100'	150'	200'	300'
1 or less	16	16	14	12	12
1-1/2	14	14	12	10	8
2	10	10	8	8	6
3	8	8	6	6	4

Eye Protection

Eye protection is required whenever there is reasonable probability of injury, and the use of suitable eye and/or face protectors is mandatory when exposed to any of the following:

Arc welding, heavy gas cutting, scarfing.
Gas welding, brazing and cutting.
Machinery or grinding of any materials causing flying particles.
Spot welding.
Powder actuated tools.
Abrasive blasting.
Injurious radiations.

6. ABOUT BUSINESS AND MANAGEMENT

Examination questions on business and management may vary greatly. They may be limited to excerpts from local statutes and codes dealing with legal financial requirements, insurance and licensing, or they may include general, management, reading an accountant's statement and/or cost estimating and job pricing. Whatever these questions deal with, we may expect more questions on business and management in forthcoming exams than ever before.

THE LEGALITY OF BUSINESS

The principal forms of business organization that may be used for a mechanical contracting business are (1) single proprietorship, (2) partnership, and (3) corporation. Each has advantages and disadvantages. Because of limited financial liability for stockholders, the corporation is the most common form of organization among contractors. Freedom of operation and tax angles are other things to be considered. Your decision about which to use should be made only after careful study and consultation with an accountant and attorney.

MANAGEMENT FUNCTIONS

Each year, several thousand businesses fail. According to the Small Business Administration, 9 out of every 10 failures are caused by poor management. Running a business requires more than knowing how to sweat a joint. Analysts regard the following as being the important areas in which poor management finally results in business failure:

Unbalanced experience and inability to;

1. Control operating expenses
2. Grant credit intelligently
3. Manage finances
4. Price jobs correctly

It is not uncommon to hear an ex-contractor blame his failure on "the competition", "a crooked bookkeeper", "lousy help", "cheap customers", but in the final analysis it is he himself who failed at his management function.

The management functions are:

1. General administration
2. Legal matters
3. Personnel
4. Design and engineering

WHAT FORM OF BUSINESS ORGANIZATION?

Individual Proprietorship

One person as sole owner.
No legal organization requirements.
Unlimited personal liability for business debts.
Termination upon death of owner.
Relative freedom from Government control.
No income tax levy on business (but on owner only).

Partnership

Two or more persons as co-owners.
Definition of partner's rights and duties by partnership agreement and State partnership law.
Possible requirement for filing certificate under State law.
Unlimited liability of each partner for all debts of firm.
Capacity of one partner to bind others when acting within scope of business.
Termination upon death or withdrawal of a partner.
Relative freedom from government control.
No income tax levy on partnership itself.

Corporation

Creation as a legal entity pursuant to State law.
Stockholders as owners who are separate and distinct from corporation.
Requirement for obtaining charter from State.
Required payment of filing fees and capital stock taxes.
Continuity unaffected by death or transfer of stock shares by any or all owners.
Limited financial liability for stockholders.
Subjection to more Government control than a proprietorship or partnership.
Income tax levy upon corporate profits and, in addition, upon dividends after they are paid to stockholders.

FIGURE 6.1

5.   Construction
6.   Purchasing and subcontracting
7.   Financial control
8.   Planning
9.   Sales
10.   Service

Scheduling and coordinating the construction department is among the most important overall responsibilities. Planning means keeping the company's resources occupied constantly without ever overexpanding beyond those resources. Those resources are obviously; manpower, equipment and finances.

When a company's sales order backlog is below what is required to fully utilize its manpower, equipment and finances, soft pricing is needed. Conversely, the higher the sales order backlog, the firmer the policy. Soft pricing means: price concessions, liberalized credit policy and terms, special discounts, dynamic sales promotions, and other inducements such as early delivery, extended warrantys and attractive service programs.

Record Keeping

No contracting business can operate profitably without accurate record keeping.

First, the Federal Government and most States require that businesses make up annual operating and financial statements for tax purposes. Then, financial control of a contracting business requires a *cash flow* picture. An appropriate *journal* or *cash book* must be kept in which every in-and-out item is recorded. The *general ledger* is divided into different accounts that may include: asset accounts, liability accounts, net worth accounts, and income and expense accounts. This information is the source of *profit-and-loss statements* (P&L) and *balance sheets*.

More than keeping records, they must be used. Constant study of company records gives management a "feel" for the business, which leads to an instant response to variations that is almost intuitive.

Job cost analysis — with accurate feedback - labor time reports, weekly payroll records, wage report and employees, are all part of the record keeping routine and the responsibility of management.

HOW TO READ A FINANCIAL STATEMENT *

Profit-and-loss statements prepared by accountants show not only dollar totals, but usually also the percentages of sales represented by each item. Percentages, of course, are expressions of arithmetical proportions. Proportions are ratios.

There are three kinds of ratios. The first are *balance sheet ratios* that refer to relationships between various balance sheet items. The second are the *operating ratios* that show the relationship of expense accounts to income. The third group is made up of ratios that show the relationship between an item in the P&L statement and one on the balance sheet.

A balance sheet tells how a business stands at one given moment in the business year. A profit-and-loss statement sums up the results of operations over a period of time.

Of themselves, these two types of financial documents are a collection of inanimate figures. But when the assorted financial symbols are interpreted and evaluated, they begin to talk.

A single balance sheet is like the opening chapters of a book — it gives the initial setting. Thus, one balance sheet will show how the capital is distributed, how much is in the various accounts, and how much surplus of assets over liabilities exists. A lone profit-and-loss statement indicates the sales volume for a given period, the amount of costs incurred, and the amount earned after allowing for all costs.

When a series of balance sheets for regularly related intervals, such as fiscal or calendar yearends, is arranged in vertical columns so that related items may be compared, the changes in these items begin to disclose trends. The comparative balance sheets then no longer remain snapshots, but are converted into X-rays, penetrating outward tissue and outlining skeletal structure of all basic management actions and decisions.

Thus, decisions to increase basic inventories because of upward price changes may be revealed in larger quantities of merchandise on hand from one period to the next. If credits are relaxed and collections slow up when sales remain constant, there may be a successive increase in receivables. If expansion is undertaken, debts may run higher; and if losses are sustained, net worth declines.

* Adapted from Ratio Analysis for Small Business, Richard Sanzo, published by The Small Business Administration, Washington, D.C., 1957.

Similarly, comparative profit-and-loss statements reveal significant changes in what took place. Were prices cut to meet competition? Then look for a lower gross profit — unless purchasing costs were reduced proportionately. Did sales go up? If so, what about expenses? Did they remain proportionate? Was more money spent on office help? Where did the money come from? How about fixed overhead? Was it controlled? It is only by comparing operating income and cost account items from one period to another that revealing answers are found.

Ratios - A Physical Exam to Determine a Company's Health

The following 10 ratios are suggested as key ones for small business purposes:

1. Current assets to current liabilities.
2. Current liabilities to tangible net worth.
3. Turnover of tangible net worth.
4. Turnover of working capital.
5. Net profits to tangible net worth.
6. Average collection period of receivables.
7. Net sales to inventory.
8. Net fixed assets to tangible net worth.
9. Total debt to tangible net worth.
10. Net profit on net sales.

Brief definitions of these ratios appear below, followed by specific examples using data taken from the balance sheet and profit-and-loss statement on (Figures 6.2, 6.3). Explanation of the terms of the financial statements, used in calculating the ratio, is also included in the discussion of each ratio.

1. *Current assets to current liabilities.* Widely known as the "current ratio," this is one test of solvency, measuring the liquid assets available to meet all debts falling due within a year's time.

Example: $\dfrac{\text{Current assets}}{\text{current liabilities}} = \dfrac{\$37,867}{\$19,242} = 1.97$ times.

Current assets are those normally expected to flow into cash in the course of a merchandising cycle. Ordinarily these include cash, notes and accounts receivable, and inventory, and at times, in addition, short term and marketable securities listed on leading exchanges at current realizable values. While some concerns may consider current items such as cash-surrender value of life insurance as current, the tendency is to post the latter as noncurrent.

Current liabilities are short term obligations for the payment of cash due on demand or within a year. Such liabilities ordinarily include notes and accounts payable for merchandise, open loans payable, short term bank loans, taxes, and accruals. Other sundry short term obligations, such as maturing equipment obligations and the like, also fall within the category of current liabilities.

2. *Current liabilities to tangible net worth.* Like the "current ratio," this is another means of evaluating financial condition by comparing what is owed to what is owned. If this ratio exceeds 80 percent, this is considered a danger sign.

Example: $\dfrac{\text{Current liabilities}}{\text{Tangible net worth}} = \dfrac{\$19,242}{\$33,970} = 56.6$ percent

Tangible net worth is the worth of a business, minus any tangible items in the assets such as goodwill, trademarks, patents, copyrights,leaseholds, treasury stock, organization expenses, or underwriting discounts and expenses. In a corporation, the tangible net worth would consist of the sum of all outstanding capital stock — preferred and common — and surplus, minus intangibles. In a partnership or proprietorship, it could comprise the capital account, or accounts, less the intangibles.

A word about "intangibles." In a going business, these items frequently have a great but undeterminable realizable value. Until these intangibles are actually liquidated by sale, it is difficult for an analyst to evaluate what they might bring. In some cases, they have no commercial value except to those who hold them: for instance, an item of goodwill. To a profitable business up for sale, the goodwill conceivably could represent the potential earning power over a period of years, and actually bring more than the assets themselves. On the other hand, another business might find itself unable to realize anything at all on goodwill.

3. *Turnover of tangible net worth.* Sometimes called "net sales to tangible net worth," this ratio shows how actively invested capital is being put to work by indicating its turnover during a period. It helps measure the profitability of the investment. Both overwork and underwork of tangible net worth are considered unhealthy.

Example: $\dfrac{\text{Net sales}}{\text{tangible net worth}} = \dfrac{\$189,754}{\$33,970} = 5.6$ times

Turnover of tangible net worth is determined by dividing the average tangible net worth into net sales for the same periods. The ratio is expressed as the number of times the turnover is obtained within the given period.

4. *Turnover of working capital.* Known, as well, as the ratio of "net sales to net working capital," this ratio also measures how actively the working cash in a business is being put to work in terms of sales. Working capital or cash is assets that can readily be converted into operating funds within a year. It does not include invested capital. A low ratio shows unprofitable use of working capital; a high one, vulnerability to creditors.

Example: $$\frac{\text{Net sales}}{\text{working capital}} = \frac{\text{net sales}}{\text{current assets-current liabilities}}$$

$$= \frac{\$189,754}{\$37,867 - \$19,242} = 10.2 \text{ times.}$$

Deduct the sum of the current liabilities from the total current assets to get working capital, the business assets which can readily be converted into operating funds. A business with $100,000 in cash, receivables, and inventories and no unpaid obligations would have $100,000 in working capital. A business with $200,000 in current assets and $100,000 in current liabilities also would have $100,000 working capital. Obviously, however, items like receivables and inventories cannot usually be liquidated overnight. Hence, most businesses require a margin of current assets over and above current liabilities to provide for stock and work-in-process inventory, and also to carry ensuing receivables after the goods are sold until the receivables are collected.

5. *Net profits to tangible net worth.* As the measure of return on investment, this is increasingly considered one of the best criteria of profitability, often the key measure of management efficiency. Profits "after taxes" are widely looked upon as the final source of payment on investment plus a source of funds available for future growth. If this "return on capital" is too low, the capital involved could be better used elsewhere.

Example: $$\frac{\text{Net profits (after taxes)}}{\text{tangible net worth}} = \frac{\$5,942}{\$33,970} = 17.5 \text{ percent.}$$

This ratio relates profits actually earned in a given length of time to the average net worth during that time. Profit here means the revenue left over from sales income and allowing for payment of all costs. These include costs of goods sold, writedowns and chargeoffs, Federal and other taxes accruing over the period covered, and whatever miscellaneous adjustments may be necessary to reduce assets to current, going values. The ratio is determined by dividing tangible net worth at a given period into net profits for a given period. The ratio is expressed as a percentage.

6. *Average collection period of receivables.* This ratio, known also as the "collection period" ratio, shows how long the money in a business is tied up in credit sales. In comparing this figure with net maturity in selling terms, many consider a collection period excessive if it is more than 10 to 15 days longer than those stated in selling terms. To get the collection period figure, get average daily credit sales, then divide into the sum of nores and accounts receivable.

Example: $\dfrac{\text{Net (credit) sales for year}}{365 \text{ days a year}}$ = daily (credit) sales ($519).

Average collection period = $\dfrac{\text{notes and accounts receivable}}{\text{daily (credit) sales}}$

$$= \dfrac{\$26,765}{\$519} = 51.6$$

This figure represents the number of days' sales tied up in trade accounts and notes receivable or the average collection received. The receivables discounted or assigned with recourse are included because they must be collected either directly by borrower, or by lender; if uncollected, they must be replaced by cash or substitute collateral. A pledge with recourse makes the borrower just as responsible for collection as though the receivables had not been assigned or discounted. Aside from this, the likely collectibility of all receivables must be analyzed, regardless of whether or not they are discounted. Hence all receivables are included in determining the average collection period.

7. *Net sales to inventory.* Known also as a "stock-to-sales" ratio, the hypothetical "average" inventory turnover figure is valued for purposes of comparing one company's performance with another, or with the industry's.

Example: $\dfrac{\text{Net sales}}{\text{Inventory}} = \dfrac{\$189,754}{\$10,385} = 18.3$ times.

A manufacturers' inventory is the sum of finished merchandise on hand, raw material, and material in process. It does not include supplies unless they are for sale. For retailers and wholesalers, it is simply the stock of salable goods on hand. It is expected that inventory will be valued conservatively on the basis of standard accounting methods of valuation, such as its cost or its market value, whichever is the lower.

Divide the average inventory into the net sales over a given period. This shows the number of times the inventory turned over in the period selected. It is compiled purely and only for purposes of making comparisons in this ratio from one period to another, or for other comparative purposes. This ratio is not an indicator of physical turnover. The only accurate way to obtain a physical turnover figure is to count each type of item in stock and compare it with the actual physical sales of that particular item.

Some people compute turnover by dividing the average inventory value at cost into the cost of goods sold for a particular period. However, this method still gives only an average. A hardware store stocking some 10,000 items might divide its dollar inventory total into cost of goods sold and come up with a physical average; this however, would hardly define the actual turnover of each item from paints to electrical supplies.

8. *Fixed assets to tangible net worth.* This ratio, which shows the relationship between investment in plant and equipment and the owners' capital, indicates how liquid is net worth. The higher this ratio, the less the owner's capital is available for use as working capital, or to meet debts.

Example: $\dfrac{\text{Fixed assets}}{\text{tangible net worth}} = \dfrac{\$15,345}{\$33,970} = 45.2$ percent.

Fixed assets means the sum of assets such as land, buildings, leasehold improvements, fixtures, furniture, machinery, tools, and equipment, less depreciation. The ratio is obtained by dividing the depreciated fixed assets by the tangible net worth.

9. *Total debt to tangible net worth.* This ratio also measures "what is owed to what is owned." As this figure approaches 100, the creditors' interest in the business assets approaches the owner's.

Example: $\dfrac{\text{Total debt}}{\text{tangible net worth}} = \dfrac{\text{current debt+fixed debt}}{\text{tangible net worth}}$

$$= \frac{\$19,242}{\$33,970} = 56.6 \text{ percent.}$$

Total debt is the sum of all obligations owed by the company such as accounts and notes payable, bonds outstanding, and mortgages payable. The ratio is obtained by dividing the total of these debts by tangible net worth.

10. *Net profit on net sales.* This ratio measures the rate of return on net sales. The resultant percentage indicates the number of cents of each sales dollar remaining, after considering all income statement items and excluding income taxes.

A slight variation of the above occurs when net operating profit is divided by net sales. This ratio reveals the profitableness of sales — i.e., the profitableness of the regular buying, manufacturing, and selling operations of a business.

To many, a high rate of return on net sales is necessary for successful operation. This view is not always sound. To evaluate properly the significance of the ratio, consideration should be given to such factors as (1) the value of sales (2) the total capital employed and (3) the turnover of inventories and receivables. A low rate of return compared with rapid turnover and large sales volume, for example, may result in satisfactory earnings.

Example: $\dfrac{\text{Net profits}}{\text{net sales}} = \dfrac{\$5,942}{\$189,754} = 3.1 \text{ percent.}$

Analyzing the Profit-and-Loss (Income) Statement

Based solely on data taken from the profit-and-loss (P&L) statement, operating ratios show the percentage relationships of each item to a common base of net sales. These percentages may be compared with those of previous periods to measure a firm's performance. They also may be compared to the typical percentages of business in similar trades or industries when they are available. Such comparisons will indicate the competitive strengths and weaknesses of a business.

The items included in profit and loss statements vary from business to business. For example, some businesses break down their sales expense to show the costs of salesmen's salaries and commissions, advertising, delivery costs, supplies, and so forth; some do not. In the following explanation of the P & L items, only major items are included.

The following explanations briefly discuss each term in the accompanying condensed profit-and-loss statement.

Net sales. This figure represents gross dollar sales minus merchandise returns and allowances. Some accountants also deduct cash discounts granted to customers on the theory that these are actually a reduction of the net selling price; others credit the discounts to "other" expense. "Trade" and "quantity" discounts are, of course, concessions off price, and should be deducted from the gross sales. In setting up the profit-and-loss statement in percentages, the net sales are shown as 100 percent.

Cost of goods sold. For retailers and wholesalers, this figure is the inventory at the beginning, plus purchases, plus "Freight in," and less inventory at the end of the period. "Freight out" is generally shown as delivery expense, either under separate or other sections of the statement.

For manufacturers, there are various additional items to be considered. They include supervision, power, supplies, the direct costs of manufacturing labor (including social security and unemployment taxes on factory employees), that portion of depreciation which enters into cost of production, and many others.

Gross profit on sales. This figure is obtained by deducting cost-of-goods sold from net sales.

Selling expenses. These expenses include such items as salaries of salesmen and sales executives, wages of other sales employees, commissions, travel expense, entertainment expense, and advertising.

Operating profit. This is the difference between the gross profit on sales and the sum of the selling expenses.

General and administrative expenses. These expenses include officers' salaries, office overhead, light, heat, communication, salaries of general office and clerical help, cost of legal and accounting services, "fringe" taxes payable on administrative personnel, sundry types of franchise and similar taxes, and other expenses.

```
                    ANY SMALL BUSINESS, INC.

                         Balance Sheet

                       December 31, 19__

                            Assets

Current assets:
  Cash on hand and in banks. . . . . . . . . . . . . $ 4,320
  Notes receivable . . . . . . . . . . . . $ 4,820
    Less notes discounted. . . . . . . . .   3,000     1,820

  Accounts receivable. . . . . . . . . . .  21,945
    Less reserve for bad debts . . . . . .   1,875    20,070

  Inventories. . . . . . . . . . . . . . . . . . . .  10,385
  Prepayment of expenses . . . . . . . . . . . . . .   1,272

        Total current assets. . . . . . . . . . .  $37,867

Plant and equipment:
  Land and building. . . . . . . . . . . $14,495
  Equipment, fixtures, and furniture . . .  4,800
    Less allowances for depreciation . . .  3,950     ·15,345

Intangibles:
  Goodwill . . . . . . . . . . . . . . . .     500
  Patent, franchises, etc. . . . . . . . .     500      1,000

        Total assets. . . . . . . . . . .          $54,212

                          Liabilities

Current liabilities:
  Notes Payable (to banks) . . . . . . . . . . . . . $ 4,000
  Accounts Payable (trade) . . . . . . . . . . . . .  10,322
  Taxes payable. . . . . . . . . . . . . . . . . . .   3,600
  Other payables (including accruals). . . . . . . .   1,320

        Total current liabilities . . . . . . . . . $19,242
Fixed liabilities. . . . . . . . . . . . . . . . . .       0

        Total liabilities . . . . . . . . . . . . . $19,242

                           Capital

Capital stock (preferred). . . . . . . . . $ 5,000
Capital stock (common) . . . . . . . . . .  20,000
Surplus. . . . . . . . . . . . . . . . . .   9,970

        Total . . . . . . . . . . . . . . . . . . . $34,970
        Total liabilities and capital . . . . . . . $54,212
```

FIGURE 6.2

```
               ANY SMALL BUSINESS, INC.

               Profit and Loss Statement

           For year ending December 31, 19___
```

Item	Amount		Percent
Gross sales. $193,472			
Less returns and allowances. .	3,718		
Net Sales.		$189,754	100.00
Cost of goods sold		147,348	77.65
Gross profit on sales.		$ 42,406	22.35
Selling expenses . . . .$ 10,479			5.52
General and administrative exp . 19,510			10.28
Financial expenses 1,312			.69
Total expenses		31,301	16.49
Operating profit		$ 11,105	5.86
Extraordinary expenses		300	.16
Net profit before taxes.		$ 10,805	5.70
Federal, State, and local taxes.		4,863	2.56
Net profit after taxes		$ 5,942	3.14

FIGURE 6.3

Financial expenses. This item would include interest, doubtful accounts, and discounts granted if not already deducted from sales.

Other operating expenses and income. Here might be included various unusual expense items not elsewhere classified, such as moving expenses, against which might be credited income from investments and miscellaneous credits and debits.

Extraordinary charges (if any). Such expenses do not occur very often, but occasionally unusual costs such as losses on sale of unused fixtures and equipment do arise.

Net profit before taxes. This figure is the profit after deducted the regular and extraordinary business charges mentioned above.

Taxes. This item includes the Federal, State and local taxes paid by the company out of its earnings.

Net profit after taxes. This figure is the final figure showing earnings available for distribution or retention.

Figure 6.3 illustrates how a condensed profit-and-loss statement would be expressed, first in terms of dollars, then in terms of percentages of net sales.

Here are some established average ratios for air conditioning, plumbing and mechanical contractors. Companies within this range may be considered healthy operations.

1. Current assets to current liabilities   2 times
2. Net profit on net sales   1.5 percent
3. Net profit on tangible net worth   10 percent
4. Net sales to tangible net worth   7.5 times
5. Current liabilities to tangible net worth   100 percent
6. Total liabilities to tangible net worth   150 percent
7. Net profits on net working capital   15 percent

COST ESTIMATING

Usually, when cost estimating questions appear on examinations, they will be part of a larger problem related to plan reading or job design. Such questions are meant to test the candidate's general understanding of how to put a job together in a business like manner, and are not concerned with detailed costing for competitive bidding.

Figure 6.4 shows a typical estimating form for pricing material and labor. After totalling all labor, material, subcontracts, and other *direct job costs,* the desired markup is then added to these costs to find the correct selling price. *Markup* is a key word in any question or problem on cost estimating. The markup is the total amount of overhead margin and profit margin; it is a different figure for each individual business. To find the markup, first determine what the percent of overhead is. The percent of overhead may only be known from your accountant's Statement of Condition. The percent overhead added to the desired percent profit (net profit) equals the total percent markup.

Example: Assume a mechanical construction cost totals $30,000.00, the percentage of overhead from the statement is 18%, the desired net profit is 10%. Find the correct selling price based on cost.

Solution: $\dfrac{\$30,000}{1} \times \dfrac{28}{100} = \$\,8,400$

$30,000 + \$8,400 = \$38,400$, the selling price

FORM E66

ESTIMATE PRICING MATERIAL & LABOR

PROJECT _____

ADDRESS _____

_____

DATE IN _____ DATE DUE _____ ESTIMATOR _____

JOB NO. _____

SHEET _____ OF _____ SHEETS

_____

_____

CHECKED BY _____

QUAN.	DESCRIPTION	LABOR		MATERIAL		EXTENSION
		UNIT	TOTAL	UNIT	TOTAL	
	MISCELLANEOUS MATERIAL AND LABOR					
	NON PRODUCTIVE LABOR					
	JOB OVERHEAD					
	SUB TOTAL: COST OF JOB					
	% GENERAL OVERHEAD					
	MATERIAL/LABOR RATIO					
	% PROFIT ON LABOR					
	% PROFIT ON MATERIAL					
	SUB CONTRACTS					
	% PROFIT ON SUBS					
	SALES TAX					
	SERVICE RESERVE					
	GRAND TOTAL: SELLING PRICE					

© 1970 Technical Guide Publications/5241 N.E. 2 Ave./Miami, Florida 33137

FIGURE 6.4

or, $30,000 x 1.28 = $38,400

In the above example the $8,400 markup is equal to 28%
profit on *cost price*. However, if the 28% profit is to be
realized on the *selling price,* then; the markup — and consequently the selling price — would be much higher.

> *Example:* Using the figures in the above example, find
> the correct selling price based on 28% profit
> on the selling price.

Solution:

> Let 100% = selling price
> 28% = marging
> 72% = cost price or $30,000

then;   100% = $\dfrac{\$30,000}{.72}$ = $41,666, the selling price,

leaving a markup of $11,666 or $3,266 more than if the profit
was based on a percentage of cost.

Table 6.1 gives conversions from % markup on cost to %
markup on selling price. Table 6.2 gives the multiplier to
price up any job based on a percentage of the selling price
where the overhead margin and profit margin have been determined.

Correct estimating and job costing procedures require
much more detailed analysis than that covered in the present
discussion. Actually, the overhead should be broken out and
assigned in separate quantities against labor and material.
The overhead assignment against labor will always be considerably greater than the overhead assignment against material.
This system *of dual overhead application rate* is based on the
average materials/labor ratio.

TABLE 6.1

PROFIT - PERCENTAGE CONVERSION

% MARKUP ON COST	% PROFIT ON SELLING PRICE	% MARKUP ON COST	% PROFIT ON SELLING PRICE
5.00	4.75	31.58	24.00
7.50	7.00	33.33	25.00
10.00	9.00	35.00	26.00
11.11	10.00	37.50	27.25
12.36	11.00	40.00	28.50
12.50	11.12	42.86	30.00
13.63	12.00	45.00	31.00
14.95	13.00	47.00	32.00
16.28	14.00	50.00	33.33
16.43	14.25	52.85	35.00
17.65	15.00	55.00	35.50
19.05	16.00	60.00	37.50
20.00	16.67	65.00	39.50
20.49	17.00	66.66	40.00
21.96	18.00	70.00	41.00
23.46	19.00	75.00	42.75
25.00	20.00	80.00	44.50
26.58	21.00	85.00	46.00
28.21	22.00	90.00	47.50
29.88	23.00	100.00	50.00

TABLE 6.2

PRECALCULATED MULTIPLIERS TO FIND SELLING PRICE

PERCENT OVERHEAD FROM STATEMENT	NET PROFIT REQUIRED BEFORE TAXES								
	4%	5%	6%	7%	8%	9%	10%	12%	15%
10	1.16	1.18	1.19	1.20	1.22	1.23	1.25	1.28	1.33
11	1.18	1.19	1.20	1.22	1.23	1.25	1.27	1.30	1.35
12	1.19	1.20	1.22	1.23	1.25	1.27	1.28	1.32	1.37
13	1.20	1.22	1.23	1.25	1.27	1.28	1.30	1.33	1.39
14	1.22	1.23	1.25	1.27	1.28	1.30	1.32	1.35	1.41
15	1.23	1.25	1.27	1.28	1.30	1.32	1.33	1.37	1.43
16	1.25	1.27	1.28	1.30	1.32	1.33	1.35	1.39	1.45
17	1.27	1.28	1.30	1.32	1.33	1.35	1.37	1.41	1.47
18	1.28	1.30	1.32	1.33	1.35	1.37	1.39	1.43	1.49
19	1.30	1.32	1.33	1.35	1.37	1.39	1.41	1.45	1.52
20	1.32	1.33	1.35	1.37	1.39	1.41	1.43	1.47	1.54
21	1.33	1.35	1.37	1.39	1.41	1.43	1.45	1.49	1.56
22	1.35	1.37	1.39	1.41	1.43	1.45	1.47	1.52	1.59
23	1.37	1.39	1.41	1.43	1.45	1.47	1.49	1.54	1.61
24	1.39	1.41	1.43	1.45	1.47	1.49	1.52	1.56	1.64
25	1.41	1.43	1.45	1.47	1.49	1.52	1.54	1.59	1.67
26	1.43	1.45	1.47	1.49	1.52	1.54	1.56	1.61	1.70
27	1.45	1.47	1.49	1.52	1.54	1.56	1.59	1.64	1.72
28	1.47	1.49	1.52	1.54	1.56	1.59	1.61	1.67	1.75
29	1.49	1.52	1.54	1.56	1.59	1.61	1.64	1.70	1.79
30	1.52	1.54	1.56	1.59	1.61	1.64	1.67	1.72	1.82

Example: Total job cost is $25,000, overhead from statement is 15%, net profit required for this job is 6%. Find profit and selling price.

Solution: From the table find the multiplier = 1.27, then; $25,000 x 1.27 = $31,750 selling price. The markup is $31,750 - 25,000 = $6,750 profit.

BIBLIOGRAPHY

John Immer, *Starting and Managing a Small Business*, Small Business Administration, Washington, D.C., 1962

Wendell Metcalf, *Starting and Managing a Small Business of Your Own*, Small Business Administration, Washington, D.C., 1962

Richard Sanzo, *Ratio Analysis for Small Business*, Small Business Administration, Washington, D.C., 3rd ed., 1970

Mark Greene, *Insurance and Risk Management for Small Business*, Small Business Administration, Washington, D.C., 2nd ed., 1970

"Your 1968 Guide to Business Management", *HAC Heating and Air Conditioning Contractor*, April 1968

Key Business Ratios in 125 Lines, Dun and Bradstreet, Business Information Systems, N.Y.

Roy Foulke, *Practical Financial Statement Analysis*, 6th ed., 1968, McGraw-Hill, N.Y.

John Gladstone, *Mechanical Estimating Guidebook*, 4th ed., 1970, McGraw-Hill, N.Y.

7.CONTRACTORS LAW

A certificate of competency to perform as a contractor
requires a legal responsibility on the part of the contractor.
In most examinations for license, a section of questions will
be devoted to such legal responsibilities.

Laws and regulations as imposed by governments are
changing instruments. Sometimes the changes are slow and
barely noticeable; at other times the changes may be acceler-
ated and broad in scope. Often, many revisions are accomplished with a change of government administration.

Contractors are usually clustered in the larger urban
communities because of the availability of material, manpower,
and communications; and of course, it is in the larger cities
that the bulk of the construction takes place. Every major
town and city, and most of the smaller communities too, has
a building code that sets forth certain requirements and
restrictions upon all new construction. Such a code is
usually state wide or regional, other codes may be national
and adopted locally by referral. Major codes were discussed
earlier in Chapter 3.

There are, of course, many other laws affecting specific
areas of building and contracting, such as construction licensing laws, health laws affecting septic tanks, wells and
potable water, elevator laws, fire laws, etc. Other laws
are state wide and affect every business and industry. These
are unemployment compensation, workmen's compensation, sales
tax and so forth. The laws are available to any interested
parties and it is your responsibility to keep abreast of any
current documents. To operate your business you must consult
with your local authorities as well as your attorney and
accountant.

OSHA

Perhaps the most important single law for contractors
is OSHA. OSHA is an acronym for The Williams-Steiger Occupational Safety and Health Act of 1970, United States Department of Labor. The purpose of this landmark legislation
is to assure safe and healthful working conditions for all
workers in the U.S.A. The Law provides that each employer
has the basic duty to furnish his employees a place of employment and a kind of employment that are free from recognized hazards to the life and health of those employees.

Although this duty is one which employers have long
accepted as a moral obligation, The Williams-Steiger Act
places specific responsibilities on each employer and

310

subjects him to serious penalties if he fails to discharge
them.   In its first fiscal year, OSHA issued over 23,000
citations with resulting penalties amounting to $2,300,000.

The Mechanical Contractors Association of America, Inc.,
has published an excellent summary of OSHA especially for the
mechanical trades.   It is called *OSHA Handbook for Mechanical
Contractors* and is reprinted here in its entirety by kind
permission of MCA.   License candidates should be thoroughly
familiar with OSHA in addition to the conditions of general
safety discussed in Chapter 5.   The contractor cannot seek
the help of his accountant or attorney in connection with job
safety and hazardous working conditions as he can with say,
tax laws, social security, unemployment compensation, etc.

OSHA   HANDBOOK

FOR

MECHANICAL   CONTRACTORS

November 1972

Second Printing
January 1973

Third Printing
April 1973

Fourth Printing
June 1973

Fifth Printing
January 1974

MECHANICAL CONTRACTORS ASSOCIATION OF AMERICA, INC.
5530 Wisconsin Ave., Washington, D. C. 20015, Suite 750
Tel. (202) 654-7960

Price: Members $0.75
       Non-Members $3.00

CONTENTS

NOTE:

1. The requirements reflected in this handbook are keyed to certain Subparts of OSHA "Safety and Health Regulations for Construction" CFR 29, PART 1926 (formerly PART 1518) with the numbered paragraph of the Regulation shown under the Subpart titles.

2. The various OSHA regulations are subject to changes by the Department of Labor at irregular intervals. All changes to regulations are published in the Federal Register. In addition, MCAA Safety Bulletins will continue to reflect pertinent revisions.

3. APPLICABILITY OF STANDARDS. The Occupational Safety and Health Administration has been queried as to whether both the OSHA Standards (PART 1910) for general industry and Construction Regulations PART 1926 (formerly 1518) apply to construction projects. OSHA's reply - "In situations where both types of standards are involved, the construction standards prevail. Where conditions are not covered by the construction, but are covered by general standards, the general ones will apply."

iv

FOREWORD

This handbook has been prepared to help MCAA members to have a better understanding of certain requirements established under the Occupational Safety and Health Act of 1970. It highlights portions of the regulations that are particularly applicable to mechanical contractors.

Section I contains a condensed version of the OSHA "Safety and Health Regulations for Construction", Title 29 CFR, PART 1926. This part was prepared by extracting and paraphrasing particularly pertinent Subparts of the regulations to reflect some of the more important requirements. It has been written so that it may be used by employees at all levels.

Section II covers some of the requirements and procedures involved with administration of OSHA that affect employers and organizational management. As in the case of Section I, this portion is an attempt to present highlights of many regulations in a consolidated and succinct form.

Many of the OSHA regulations, requirements and details have been omitted in order to present material in an abridged edition. Consequently, this handbook is not, nor is it intended to be a substitute for, or a book for use in lieu of, the Act or unabridged Federal and State regulations and requirements, which must be followed to assure precise compliance with OSHA. It is intended, however, to help make OSHA more understandable and to assist MCAA members to improve their safety programs and practices.

The material in this handbook is supplemented by other MCAA publications on safety, such as the series of MCAA Safety Committee Bulletins. These publications will continue to keep our membership informed on safety matters. It is our hope that these combined efforts will enhance safety throughout our industry.

The MCAA SAFETY COMMITTEE

James G. Hendrickson

C. T. Hammond

H. R. Patterson

SECTION I

SAFETY AND HEALTH REGULATIONS FOR CONSTRUCTION

GENERAL SAFETY AND HEALTH PROVISIONS
(Reference: SUBPART C, Paragraphs 1926.20-.32)

CONTRACTOR
RESPONSI-
BILITY

1. Contractor Requirements - No contractor or subcontractor shall require any employee to work in surroundings or under working conditions which are unsanitary, hazardous or dangerous to his work or safety.

2. It is the responsibility of the employer to initiate and maintain programs for compliance with OSHA regulations. Such programs shall provide for regular and frequent inspections of sites, materials and equipment by competent persons designated by the employer.

IDENTIFY
UNSAFE
EQUIPMENT

3. The use of material or equipment not in compliance with OSHA regulations is prohibited. Unsafe material and equipment shall be identified as unsafe by tagging, or locking controls to make it inoperative, or by removing it from its place of operation.

TRAINING

4. Only those employees qualified by training or experience shall be permitted to operate equipment and machinery.

5. The employer shall instruct each employee in safety practices to include recognition and avoidance of unsafe conditions and potential hazards and use of safety equipment. Employees required to handle flammable, toxic or otherwise dangerous materials shall be instructed in safe handling procedures and for appropriate personal hygiene and first aid measures to be used in event of injury.

MEDICAL

6. First aid services and provisions for medical care shall be made available by the employer (See Subpart D).

FIRE
PROTECTION

7. The employer shall maintain an effective fire prevention and protection program throughout all phases of construction (Subpart F).

HOUSE-
KEEPING

8. a. During construction work areas, passageways and stairs in and around the site must be kept clear of debris. Containers shall be provided for collection and separation of waste and trash.

   b. Scrap, debris, garbage and other waste shall be disposed of at regular and frequent intervals.

I-1

LIGHTING 9. All work areas shall be properly lighted with natural or artificial illumination (Subpart D).

PERSONAL
PROTECTIVE
EQUIPMENT
10. The employer is responsible for requiring the wearing of proper personal protective equipment in all operations where hazardous conditions warrant such equipment (Subpart E).

BOILERS
AND
PRESSURE
VESSELS
11. Current and valid certification by an insurance company or regulatory authority are required, attesting to safe installation inspection and testing of boilers and pressure vessels.

CONCENSUS
STANDARDS
12. Specifications, standards and codes (such as American National Standards Institute Standards) incorporated by reference in OSHA standards shall have the same force and effect as the other OSHA regulations.

Occupational Health and Environmental Controls
(Reference: SUBPART D, Paragraph 1926.50-.58)

MEDICAL
SERVICES
AND FIRST
AID

1. Medical personnel must be available for advice and consultation on occupational health.

2. At Each Worksite:
   a. Arrangements must be made in advance of a job for prompt medical attention for serious injuries. Telephone numbers of doctor(s), hospital(s) and ambulance(s) must be posted conspicuously.
   b. Unless a physician or medical facility is reasonably accessible for treatment of injuries, a person must be present having a valid certificate in First Aid from an American Red Cross or equivalent training course.

   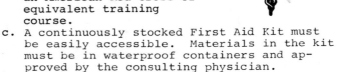

   c. A continuously stocked First Aid Kit must be easily accessible. Materials in the kit must be in waterproof containers and approved by the consulting physician.

SANITATION

3. Water
   a. Potable water shall be provided at all sites.
   b. Water containers must be closed, equipped with a tap, and clearly marked.
   c. A common drinking cup, and dipping is prohibited. Unused disposable cups must be in a sanitary container; and a receptacle provided for used cups.
   d. Non-potable water must be clearly labeled as such.

4. Toilets
   a. Toilets must be provided as follows:
      1 Toilet for 20 employees or less.
      1 Toilet seat and 1 urinal per forty workers,for 20 or more employees.
   b. Toilet rooms must be equipped with self closing doors, latch, adequate toilet paper, toilet paper holder, and clothing hooks.

OCCUPA-
TIONAL
NOISE
EXPOSURE

5. Employees must be protected against exposure to injurious sound levels by controlling exposure through feasible administrative or engineering measures. If these controls are inadequate, personal protective equipment must be provided. Cotton is unacceptable as

I-3

ear protective material. Noise exposure
level must not exceed 90 decibels per 8 hour
day. (See Exposure Table D-2, Paragraph
1926.52 in full text of "Occupational Safety
and Health Regulations" for specific noise
levels).

RADIATION   6. Employees must be protected against exposure
to ionizing (X-ray, radioactive) and non-
ionizing (laser beam) radiation.

GASES,   7. Protection against exposure to harmful gases,
VAPORS,   fumes, dusts and similar airborne hazards
FUMES,   must be furnished through engineering
DUSTS AND   (ventilation), administration (exposure time)
MISTS   or personal protective equipment (respiratory
                devices).

ILLUMI-   8. Construction and other working areas must be
NATION   adequately lighted at all time while work is
                in progress as follows:
                5 Footcandles - General construction areas,
                loading platforms, warehouses, corridors.
                10 Footcandles - Construction shops.
                30 Footcandles - Offices.

VENTILATION 9. When hazardous substances, such as dusts,
                vapors, or gases, exist in construction work
                and ventilation is used as a control, the
                system must be properly designed, installed
                and operated continually during all operations
                which it is designed to serve.

I-4

PERSONAL PROTECTIVE AND LIFE SAVING EQUIPMENT
(Reference: SUBPART E, Paragraph 1926.100-.107)

HEAD
PROTECTION

1. Employees working in areas where there is a possible danger of exposure to impact or penetration of falling or flying objects or high voltage electrical shocks and burns shall be required to wear approved plastic safety helmets (hard hats)   and required to wear eye and face protective equipment while engaged in operations that present potential eye or face injuries.

RESPIRATORY
DEVICES

2. In emergencies and when controls required for proper ventilation are inadequate, approved respiratory devices shall be provided by the employer. The employee shall be instructed in its proper care and use and shall use it.

SAFETY
BELTS,
LIFELINES
AND
LANYARDS

3. During certain operations, safety belts, lifelines, lanyards, or life preservers may be required. Safety nets are required when workplaces are more than 25 feet above the ground, water, or other surfaces where use of scaffolds, catch platforms, temporary floors or ladders are impractical.

FIRE PROTECTION AND PREVENTION
(Reference: SUBPART F, Paragraph 1926.150-.155)

FIRE PROTECTION

1. The employer is responsible for the development of a proper fire protective program; he shall provide for the firefighting equipment as specified in this subpart.

2. All firefighting equipment shall be conspicuously located, accessible, inspected periodically, maintained at all times and replaced immediately when defective.

3. A fire extinguisher rated 2A (or 55 gallon drum with 2 fire pails) shall be provided for each 3000 square feet of the protected building.

4. On multi-story buildings, at least one fire extinguisher shall be located adjacent to each stairway.

5. A fire alarm system (telephone, siren or similar device) shall be established by the employer whereby employees and the local fire department can be alerted in case of fire. Telephone number of local fire department must be conspicuously posted.

6. For ratings and types of fire extinguishers see Table F-1, Paragraph 1926.150.

FIRE EXTINGUISHERS

	WATER TYPE				FOAM	CARBON DIOXIDE	DRY CHEMICAL SODIUM OR POTASSIUM BICARBONATE		MULTI-PURPOSE ABC	
	STORED PRESSURE	CARTRIDGE OPERATED	WATER PUMP TANK	SODA ACID	FOAM	CO2	CARTRIDGE OPERATED	STORED PRESSURE	STORED PRESSURE	CARTRIDGE OPERATED
CLASS A FIRES — WOOD, PAPER, TRASH HAVING GLOWING EMBERS	YES	YES	YES	YES	YES	NO (BUT WILL CONTROL SMALL SURFACE FIRES)	NO (BUT WILL CONTROL SMALL SURFACE FIRES)	NO (BUT WILL CONTROL SMALL SURFACE FIRES)	YES	YES
CLASS B FIRES — FLAMMABLE LIQUIDS, GASOLINE, OIL, PAINTS, GREASE, ETC.	NO	NO	NO	NO	YES	YES	YES	YES	YES	YES
CLASS C FIRES — ELECTRICAL EQUIPMENT	NO	NO	NO	NO	NO	YES	YES	YES	YES	YES
CLASS D FIRES — COMBUSTIBLE METALS	SPECIAL EXTINGUISHING AGENTS APPROVED BY RECOGNIZED TESTING LABORATORIES									

FIRE PREVENTION

7. Internal combustion engines shall be located so that the exhausts are well away from combustible material.

8. Smoking shall be prohibited in vicinity of potential fire hazards. Fire hazard areas shall be conspicuously posted "No Smoking or Open Flame".

I-6

9. Temporary Buildings
   a. Interior temporary buildings shall be of
      non-combustible materials or of combus-
      tible construction having a fire resist-
      ance of not less than one hour.
   b. Exterior temporary buildings shall be
      located at least 10 feet from another
      building or structure. Groups of tem-
      porary buildings not exceeding 2000
      square feet shall be considered a single
      temporary building.

10. Open Yard Storage
   a. Combustible materials must be stacked in
      stable piles, in no case higher than 20
      feet.
   b. Driveways between and around storage piles
      shall be at least 15 feet wide. The
      entire storage site shall be kept free of
      rubbish and accumulation of unnecessary
      combustible materials, including weeds.
   c. No combustible material shall be stored
      outdoors within 10 feet of a building or
      structure.
   d. Appropriate portable fire extinguishing
      equipment shall be provided at accessible
      locations in the storage area.

11. Interior Storage
   a. Storage shall not adversely affect means
      of exit.
   b. Clearance shall be maintained around
      lights and heating units to prevent ig-
      nition of combustible materials.
   c. A barricade, or a clearance of at least
      two feet, must be maintained around path
      of a fire door. No material shall be
      stored within 3 feet of a fire door
      opening.

FLAMMABLE
AND
COMBUSTIBLE
LIQUIDS

12. Only approved containers and portable tanks
    shall be used for storage and handling of
    flammable and combustible liquids.

13. Flammable and combustible liquids shall not
    be stored in exits, stairways or other per-
    sonnel passages.

14. Indoor Storage
   a. No more than 25 gallons of flammable or
      combustible liquids shall be stored in a
      room outside of an approved storage
      cabinet.

I-7

b. Flammable and combustible liquids in excess of 25 gallons shall be stored in
acceptable or approved wood or metal cabinets (see Paragraph 1926.152 b(2) for
specifications).
c. Cabinets must be conspicuously labeled
"Flammable--Keep Fire Away". Not more
than 60 gallons of flammable liquid or
120 gallons of combustible liquids shall
be stored in any one cabinet.

15. Outdoor Storage
a. Storage of containers (not more than 60
gallons each) shall not exceed 1,100
gallons in any one pile or area. Piles
shall be separated by 5 feet and shall
not be nearer than 20 feet to a building.
b. The storage area shall be graded and
equipped with curbs, dikes and other
measures to control or direct spillage
from buildings.
c. Portable tanks shall not be nearer than
20 feet from any building.

FIRE
CONTROL

16. At least one portable fire extinguisher
rated 20-B units or more shall be located
within 10 feet of the door opening outside
any room used for storage of 60 gallons or
more of flammable or combustible liquids.

17. At least one portable fire extinguisher
rated 20-B units or more shall be located
between 25 and 75 feet from any outside
flammable liquid storage area.

SERVICE
AND
REFUELING
AREAS

18. In service and refueling areas, flammable or
combustible liquids shall be stored in approved closed containers, portable tanks, or
underground tanks.

19. No smoking or open flames shall be permitted
in fueling and servicing areas. "No Smoking"
and "No Open Flame" signs shall be posted
conspicuously.

20. Motors of equipment being fueled shall be
shut off during fueling operations.

21. Each service or fuel area shall be provided
with at least one 20 BC fire extinguisher.

LIQUIFIED
PETROLEUM
GAS

22. Every container and vaporizer shall be provided with approved safety relief valves or
devices.

23. LP-Gas consuming appliances must be approved
types.

I-8

24. Storage of LP-Gas within buildings is prohibited. Containers shall be kept upright
on firm foundations or otherwise secured.

TEMPORARY
HEATING
DEVICES

25. Ventilation shall be provided to supply
fresh air to maintain health and safety of
workers.

26. Solid fuel salamanders are prohibited in
buildings and on scaffolds.

27. Flammable liquid-fired heaters shall be
equipped with a primary safety control
(not barometric or gravity feed type) to
stop the flow of fuel in the event of flame
failure.

SIGNS, SIGNALS AND BARRICADES
(Reference: SUBPART G, Paragraph 1926.100-.103)

ACCIDENT
PREVENTION
SIGNS

1. Approved danger signs shall be used when and only when an immediate danger exists.

2. Approved caution signs shall be used when and only when a potential hazard exists or to caution against unsafe practices.

TAGS

3. Accident prevention tags shall be used as a temporary means of warning employees of an existing hazard, such as defective tools or equipment. They shall not be used in place of, or as a substitute for, accident prevention signs.

SIGNALING

4. When operations are such that signs, signals, and barricades do not provide the necessary protection on or adjacent to a highway or street, flagmen or other appropriate traffic controls shall be provided.

ACCIDENT PREVENTION TAGS

White tag - White letters on red square

White tag - White letters on red oval with a black square

Yellow tag - Yellow letters on a black background

White tag - White letters on black background

MATERIALS HANDLING, STORAGE, USE AND DISPOSAL
(Reference: SUBPART H, Paragraph 1926.250-.252)

MATERIALS STORAGE

1. Aisles and passageways shall be kept clear for free and safe movement of equipment and employees.

2. Material stored inside buildings under construction shall not be placed within 6 feet of any hoistway, or inside floor opening, or within 10' of an unfinished exterior wall.

3. Pipe, structural steel, and other cylindrical material, unless racked, shall be stacked and blocked so as to prevent spreading or tilting.

RIGGING EQUIPMENT

4. Rigging equipment shall be inspected prior to use on each shift and, when not in use, shall be removed from the immediate work area, so as not to present hazards to employees.

5. Special custom design grabs, hooks, and clamps, shall be marked to indicate the safe working loads and shall be proof tested prior to use to 125% of their rated load.

6. Alloy steel chain slings shall have permanent affixed durable identification stating size, grade, rated capacity, and manufacture.

7. When using wire rope, a safety factor of not less than 5 shall be maintained.

8. Each synthetic web sling shall be marked to show name or trade mark of manufacturer, rated capacity for the type of hitch, and type of material.

9. See Tables H-1 through H-19, Paragraph 1926.250 for specifications on rigging equipment.

DISPOSAL OF WASTE MATERIALS

10. Whenever materials are dropped more than 20' outside of a building, an enclosed chute must be used.

11. When materials are dropped through holes in the floor without the use of chutes, the area must be barricaded.

12. Waste material shall be removed from the immediate work area, as the work progresses.

13. All solvent waste, oily rags, and flammable liquids, shall be kept in fire resistant covered containers until removed from worksite.

I-11

TOOLS - HAND AND POWER
(Reference: SUBPART I, Paragraph 1926.300-.305)

GENERAL REQUIRE-MENTS

1. All hand and power tools and similar equipment, whether furnished by the employer or employee, shall be maintained in a safe condition.

2. Belts, gears, shafts, pulleys, sprockets, spindles, drums, fly wheels, chains, or other reciprocating, rotating or moving parts of such equipment shall be guarded if such parts are exposed to contact by employees or otherwise create a hazard.

3. Switches:
   a. All hand-held powered platen sanders, grinders with wheels 2-inch diameter or less, routers, planers, laminate trimmers, nibblers, shears, scroll saws, and jigsaws with blade shanks one-fourth of an inch wide or less may be equipped with only a positive "on-off" control.
   b. All hand-held powered drills, tappers, fastener, drivers, horizontal, vertical, and angle grinders with wheels greater than 2 inches in diameter, disc sanders, belt sanders, reciprocating saws, saber saws, and other similar operating powered tools shall be equipped with a momentary contact "on-off" control and may have a lock-on control provided that turnoff can be accomplished by a single motion of the same finger or fingers that turn it on.
   c. All other hand-held powered tools, such as circular saws, chain saws, and percussion tools without positive accessory holding means, shall be equipped with a constant pressure switch that will shut off the power when the pressure is released.
   d. Exception: The above does not apply to concrete vibrators, concrete breakers, powered tampers, jack hammers, rock drills, and similar hand operated power tools.

HAND TOOLS

4. Impact tools, such as drift pins, wedges and chisels, shall be kept free of mushroom heads.

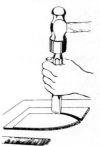

I-12

POWER
OPERATED
HAND
TOOLS

5. All electric power operated tools shall
   either be of the approved double-insulated
   type or properly grounded and have cords
   free from defects.

6. The use of hoses or electric cords for hoisting or lowering shall not be permitted.

7. Pneumatic power tools shall be secured to
   hose or whip by some positive means to prevent the tool from becoming accidentally
   disconnected.

8. Safety clips shall be securely installed on
   pneumatic impact tools to prevent attachments from being accidentally expelled.

9. Compressed air shall not be used for cleaning purposes except where reduced to less
   than 30 p.s.i. and then only with effective
   chip guarding and personal protection equipment. The 30 p.s.i. requirement does not
   apply for concrete form, mill scale and
   similar cleaning purposes.

10. All hoses exceeding 1/2-inch inside diameter
    shall have a safety device at the source of
    supply or branch line to reduce pressure
    in case of hose failure.

11. All fuel powered tools shall be stopped
    while being refueled, serviced or maintained.

12. Only employees who have been trained in the
    operation of the particular tool in use shall
    be allowed to operate a **powder-actuated** tool.

ABRASIVE
WHEELS
AND TOOLS

13. All abrasive wheels shall be provided with
    safety guards.

JACKS-LEVER
AND RATCHET,
SCREW AND
HYDRAULIC

14. All jacks shall have a positive stop to
    prevent overtravel.

WELDING AND CUTTING

(Reference: SUBPART J, Paragraph 1926.350-.354)

GAS
WELDING
AND
CUTTING

1. When transporting gas cylinders, they shall be secured on a cradle, slingboard, or pallet. Choker slings or electric magnets shall not be used for this purpose.

2. Valve protection caps shall be in place and secured, when transporting or storing gas cylinders.

3. Cylinders being transported by powered vehicles shall be secured in a vertical position, with the valve protective caps in place.

4. Unless cylinders are firmly secured on a special carrier (oxygen-acetylene cart) intended for this purpose, regulators shall be removed and valve protection caps put in place before cylinders are moved.

5. A suitable cylinder truck, chain or other steadying device shall be used to keep cylinders from being knocked over while in use.

6. The cylinder valve shall be opened only when work is being performed.

7. Gas cylinders shall be secured in an upright position at all times except, for times when cylinders are being hoisted or carried.

8. Fuel gas hose and oxygen hose shall be easily distinguishable from each other.

9. When parallel sections of oxygen and fuel hose are taped together, not more than 4 inches out of 12 inches shall be covered by tape.

10. Defective hose or hose in doubtful condition shall not be used.

11. Torches in use shall be inspected each day for leaking shut-off valves, hose couplings, and tip connections.

12. Defective torches shall not be used.

13. Torches shall be lighted by friction lighters and not by matches or from hot work.

14. Oxygen and fuel gas pressure regulators, including their related gauges, shall be in proper working order.

15. Oxygen cylinders and fittings shall be kept away from oil or grease.

ARC
WELDING
AND
CUTTING

16. Only manual electrode holders which are specifically designed for arc welding and cutting shall be used.

17. Any current-carrying parts shall be fully insulated against the maximum voltage encountered to ground.

18. All arc welding and cutting cables shall be capable of handling the maximum current requirements of the work in progress.

19. Only cables with standard insulated connectors of a capacity at least equivalent to that of the cable shall be used.

20. Cables needing repair shall not be used.

21. A ground return cable shall have a safe current carrying capacity equal to or exceeding the specified maximum output capacity of the arc welding units which it services.

22. The frames of all arc welding machines shall be grounded either through a third wire in the cable containing the circuit conductor or through a separate wire, which is grounded at the source of the current.

23. All ground connections shall be inspected to ensure that they are mechanically strong and electrically adequate for the required current.

FIRE
PREVENTION

24. When practical, objects to be welded, cut or heated shall be moved to a designated safe location; or if the objects to be welded, cut or heated cannot be readily moved, all

movable fire hazards in the vicinity must be
taken to a safe place or otherwise protected.

25. Suitable fire extinguishing equipment shall
be made available when welding, cutting, or
heating is being performed.

26. For the elimination of possible fire hazards
in enclosed spaces as a result of gas escaping through leaking or improperly closed
torch valves, the gas supply to the torch
shall be positively shut off at some point
outside the enclosed space whenever the
torch is not to be used. Overnight, the
torch and hose shall be removed from the
confined space.

27. Before heat is applied to a drum, container,
or hollow structure, a vent or opening shall
be provided for the release of any built-up
pressure during the application of heat.

ELECTRICAL
(Reference: SUBPART K, Paragraph 1926.400-.405)

GENERAL
REQUIRE-
MENTS

1. No employer shall permit an employee to work in such proximity to any part of an electric power circuit that he may contact the same in the course of his work, unless the employee is protected against electric shock by deenergizing the circuit and grounding it or by guarding it by effective insulation or other means.

2. In work areas where the exact location of underground electric power lines is unknown, workmen using jackhammers, bars, or other hand tools which may contact a line shall be provided with insulated protective gloves.

GROUNDING
AND
BONDING

3. Portable tools and appliances protected by an approved system of double insulation, or its equivalent, need not be grounded. Where such an approved system is employed, the equipment shall be distinctly marked.

4. Extension cords used with portable electric tools and appliances shall be of three-wire type.

5. Temporary lights shall be equipped with guards to prevent accidental contact with the bulb, except that guards are not required when the construction of the reflector is such that the bulb is deeply recessed.

6. Temporary lights shall be equipped with heavy duty electric cords with connections and insulation maintained in safe condition. Temporary lights shall not be suspended by their electric cords unless cords and lights are designed for this means of suspension. Splices shall have insulation equal to that of the cable.

7. Portable electric lighting used in moist and/or other hazardous locations, as for example, drums, tanks, and vessels shall be operated at a maximum of 12 volts.

EQUIPMENT
INSTALLATIONS
AND
MAINTENANCE

8. Receptacles for attachment plugs shall be of approved, concealed contact type with a contact for extending ground continuity and shall be so designed and constructed that the plug may be pulled out without leaving any live parts exposed to accidental contact.

9. Handlamps of the portable type shall be of the molded composition or other type approved for the purpose. Brass-shell, paper-lined lampholders shall not be used. Handlamps shall be equipped with a handle and a substantial guard over the bulb and attached to the lampholder or the handle.

10. Extension cords shall be protected against accidental damage as may be caused by traffic, sharp corners, or projections and pinching in doors or elsewhere.

11. Extension cords shall not be fastened with staples, hung from nails, or suspended by wire.

LADDERS AND SCAFFOLDING
(Reference: SUBPART L, Paragraph 1926.450-.451)

LADDERS

1. The use of ladders with broken or missing rungs or steps, broken or split side rails, or otherwise defective is prohibited.

2. Ladders shall not be painted.

3. Portable ladders shall be placed on a substantial base and the area around the top and bottom shall be kept clear.

4. Do not use ladders at a pitch of over one (1) to four (4). (Distance from base of ladder to the wall 1/4 the length of the ladder).

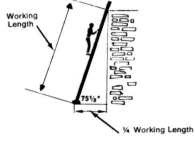

5. Ladders shall not be used in passageways or driveways unless protected by barricades.

6. Ladders shall extend not less than 36 inches above the landing unless grab rails are installed.

7. Portable ladders shall be tied, blocked or otherwise secured to prevent their being displaced.

8. Portable ladders shall be equipped with safety shoes. (Note: Portable/straight ladders without safety shoes are cited frequently by OSHA inspectors.)

9. If job made ladders are necessary, check OSHA regulations for specifications (Subpart L).

SCAFFOLDING

General Requirements:
10. All scaffolding must have solid footing or anchoring capable of holding the intended load without settling or shifting. No unstable object, as barrels, blocks, or boxes shall be used to support scaffolds or planks.

11. Guardrails and toeboards must be used on all open sides and ends of platforms which are

I-19

over 10 feet above the ground or floor. Scaffolds 4 to 10 feet high, which are less than 45 inches wide, must also have standard guardrails.

12. Guardrails must be 2 x 4 inches, or equivalent, 42 inches high with a midrail, when required. Toeboards must be at least 4 inches high and within 1/4" of the platform.

13. When persons are required to work or pass under the scaffold, an 18" gauge, or equivalent, screen must extend from the toeboard to guardrail.

14. No space over 1/2 inch wide is allowed between the planks of a platform.

15. Scaffold must be able to support four times the maximum intended load.

16. Planks must extend over the end supports, not less than six inches, or more than twelve inches.

TUBE AND COUPLER SCAFFOLDS

17. All members must be 2 inches outside diameter.

18. The posts cannot be over 6 feet apart by 10 feet along the length.

19. The scaffold must be tied to the structure at intervals of 30 feet, horizontally, by 26 feet vertically.

20. Guardrails, toeboards and wire screen must be used, as described previously.

TUBULAR WELDED FRAME SCAFFOLDS

21. Legs must set on adequate supports.

22. The frame must be aligned with coupling or stack pins.

23. The scaffolding must be tied to the building at intervals of 30 feet horizontally and 26 feet vertically.

24. Guardrails, toeboards and wire screen must be used as described previously.

MOBILE SCAFFOLDS

25. The height shall not exceed four times the minimum base dimension.

26. Casters must have positive locking devices.

27. Although there are specific exceptions, in general no one may ride a moving scaffold.

I-20

Floor and Wall Openings
(Reference: SUBPART M, Paragraph 1926.500-.502)

FLOOR OPENINGS

1. Floor openings shall be guarded by a standard railing and toeboard on all exposed sides except regular entrances to stairways.

2. Hatchways, ladderways, and skylight openings shall be guarded by standard railings or covers.

3. Pits, trap-doors, and manholes shall have standard covers.

4. All temporary floor openings must have standard railings.

WALL OPENINGS

5. Wall openings which have the bottom of the opening less than 3 ft. above the floor level and have a drop of 4 ft. must have a standard and intermediate railing. If the bottom of openings is less than 4 in. above the floor, it shall be protected by a standard toeboard.

6. Every openside floor or platform with a drop of 6 ft. or more shall be guarded by a standard railing, except where there is entrance to a ramp, stairway, or fixed ladder.

7. Regardless of height, open-sided floors, walkways, or platforms above or adjacent to dangerous equipment shall be guarded with a standard railing and toeboard.

8. Every flight of stairs having four or more risers shall be equipped with standard stair railing or handrails as specified in regulations.

STANDARD RAILING

9. A standard railing consists of a top rail, intermediate rail, toeboard and posts approximately 42 in. from surface of platform to top of rail. Meeting the following requirements:
   a. For wood railings, the posts shall be of at least 2 in. x 4 in. stock spaced not to exceed 8 ft.; the top rail shall be of at least 2 in. x 4 in. stock, the intermediate rail shall be of at least 1 in. x 6 in. stock.
   b. For pipe railings, posts, top, and intermediate railings shall be at least 1 1/2 in. nominal diameter with parts spaced not

more than 8 ft. on centers.
c. The anchoring of the railings or parts
shall be strong enough to withstand a
force of 200 pounds applied in any direction with minimal deflection.

10. A standard toeboard shall be 4 in. minimum
in vertical height from its top edge to the
level of the floor. It shall be securely
fastened in place and with not over 1/4 in.
between the floor and the bottom of the toeboard.

CRANES, DERRICKS, HOISTS, ELEVATORS, AND CONVEYORS
(Reference: SUBPART N, Paragraph 1926.550-.555)

GENERAL REQUIRE-MENTS

1. The manufacturer's name and specifications applicable to the operation of all cranes and derricks shall be posted or attached to the equipment.

2. Rated load capacities and recommended rules of operation shall be conspicuously posted on all equipment at the operator's station.

3. Hand signals to crane and derrick operators, prescribed by ANSI standards for the type of equipment in use, shall be posted at the jobsite.

4. A competent person shall inspect all machinery and equipment prior to each use, and during use to make sure it is in safe operating condition.

5. A thorough, annual inspection of hoisting machinery shall be made by a competent person.

6. Wire rope shall be taken out of service when one of the following conditions exist.
   a. In running ropes, 6 random distributed broken wires in one lay or 3 broken wires in one stand or one lay.
   b. Wear of one-third the original diameter or outside individual wires.
   c. Kinking, crushing, hoist caging, heat damage, or any other damage resulting in distortion of the rope structure.
   d. In standing ropes, more than two broken wires in one lay in sections beyond end connections, or more than one broken wire at an end connection.

7. All belts, gears, shafts, etc., shall be guarded if such parts are exposed to contact by employees.

8. All exhaust pipes shall be guarded or insulated in areas where exposed to employee contact.

9. An accessible fire extinguisher of 5 BC rating or higher shall be available at all operator stations.

OVERHEAD HOISTS

10. The safe working load of the overhead hoist

I-23

as determined by the manufacturer, shall be indicated on the overhead hoist, and this safe working load shall not be exceeded.

11. The supporting structure to which the hoist is attached shall have a safe working load equal to that of the hoist.

MOTOR VEHICLES AND MECHANIZED EQUIPMENT
(Reference: SUBPART O, Paragraph 1926.600-.606)

EQUIPMENT-
GENERAL
REQUIRE-
MENTS

1. All equipment left unattended at night adjacent to highways or construction areas
shall have lights, reflectors or barricades
with lights or reflectors to identify
location of the equipment.

2. A safety rack or cage or equivalent protection shall be used when inflating, mounting,
or dismounting tires installed on split
rims or rims with locking rings or similar
devices.

3. Heavy machinery, equipment or parts of such
equipment which are suspended or held aloft
must be lowered, blocked or cribbed to
prevent falling or shifting before employees
are allowed to work under them. Bulldozer
blades, end loader buckets, dump bodies and
similar equipment shall be lowered or blocked
when not in use or when being repaired.

4. When vehicles or mobile equipment are stopped
or parked, parking brakes shall be set. Equipment on inclines shall have wheels chocked
as well as having parking brakes set.

5. All cab glass must be safety glass or equivalent.

MOTOR
VEHICLES

(Paragraphs 6-11 apply to _motor_ vehicles
used on job site work. Requirements for
materials handling equipment are covered
in Paragraphs 12-17 below.)

6. All vehicles shall have a service brake system, parking brake system and emergency brake
system in operable condition.

7. All vehicles or combination of vehicles shall
have in operable conditions at least:
a. 2 Headlights
b. 2 Taillights
c. Brakelights
d. Audible warning device at operator's
station
e. Seat belts properly installed
f. Seats, firmly secured, for the number of
persons carried

8. No employer shall use motor equipment having
obstructed rear view unless:

a. Vehicle has an audible reverse signal alarm.
b. Vehicle is backed up only when observer says it is safe to do so.

9. All vehicles with cabs shall have:
   a. Windshields and powered wipers
   b. Cracked and broken glass replaced
   c. Defogging/defrosting devices when fog/frost may occur

10. All rubber tire vehicles shall be equipped with fenders. Mud flaps may be used in lieu of fenders when equipment is not designed for fenders.

11. All vehicles must be checked thoroughly at the beginning of each shift to assure that the vehicle is in safe operating condition. All defects must be corrected before vehicle is placed in service.

MATERIAL HANDLING EQUIPMENT

12. Earth moving equipment must be provided with brakes and audible alarms. Other safety devices, such as fenders and seat belts, are required on certain types of equipment. (See Note.)

    NOTE: Specific details on devices required are in SUBPART O, Paragraphs 1926.601 - .605 and SUBPART W - "Rollover Protective Structures; Overhead Protection," Paragraph 1926.1000 - .1003.

13. All bidirectional machines such as rollers, compactors and bulldozers must be equipped with a horn, which shall be operated as needed when the machine is moving in either direction.

14. Roadway grades for construction equipment must be designed by a qualified engineer.

15. Scissor points on all front end loaders, which are a hazard to the operation during normal operations, shall be guarded.

16. Lifting and hauling equipment shall have rated capacity posted on the vehicle so as to be clearly visible to the operator.

17. High lift rider industrial trucks shall be equipped with overhead guards.

I-26

Excavation, Trenching and Shoring
(Reference: SUBPART P, Paragraph 1926.650-.652)

GENERAL
REQUIRE-
MENTS

1. Walkways, runways and sidewalks shall be kept clear of excavated material or other obstructions and no sidewalk shall be undermined unless shored.

2. No person shall be allowed under loads handled by power shovels, derricks or hoists.

3. Before any excavation, underground utilities must be located and trees and surface objects that might create a hazard must be removed.

TRENCHING

4. Trenches over 5 feet deep in hard or compact soil shall be shored or sloped to the angle of repose of the soil

5. Trenches less than 5 feet deep in ground with hazardous conditions must be treated as above.

6. Portable trench boxes or sliding trench shields may be used in lieu of shoring or sloping.

7. Excavated material must be kept 2 feet, or more, from the trench.

8. When employees are required to be in trenches 4 feet deep, or more, ladders have to be provided so that a means of exit is available without more than 25 feet of lateral travel.

9. All excavation slopes, shoring and bracing must be inspected daily, and after each rainstorm.

SECTION II

GENERAL INFORMATION ON OSHA

GENERAL INFORMATION ON OSHA

PURPOSE   The purpose of the Occupational Safety and Health Act of 1970 (OSHA) is to assure so far as possible every working man and woman in the Nation safe and healthful working conditions and to preserve human resources.

APPLICATION   OSHA provisions apply to every employer engaged in a business affecting commerce who has employees.

RESPONSI-
BILITY   Employers - Each employer must furnish each of his employees employment and place of employment free from hazards likely to cause death or serious physical harm, and must comply with OSHA regulations. Provisions of the Act applicable to employers are enforceable by fines and other penalties.

Employees - Each employee must comply with all OSHA regulations which are applicable to his own actions.

NOTICES   1. OSHA Notices
Employers must post and keep posted notices provided by OSHA. The OSHA informational notice "Occupational Safety and Health on the Job" must be posted in each establishment in a conspicuous place or places where notices are customarily posted. Each activity, such as construction shacks, shall be treated as separate establishments.
OSHA notices must not be defaced, altered or covered with other material.
The "Occupational Safety and Health On The Job" notice may be obtained from the nearest OSHA Regional Office. (It is also available, printed on tough, durable DuPont TYVEK, from MCAA (Safety Poster 13-11) for $ .20 each with 25% discount on orders of 25 or more).
Penalty - Failure to post the OSHA informational notice subjects employers to a proposed penalty of $50.00, which may not be adjusted or abated.

2. Posting Citations
Upon receipt of any OSHA citation, employers must immediately post a copy of the citation at or near each place an alleged violation occurred.
Penalty - Failure to post citations subject employers to a proposed penalty of $500.00 (each citation), which may not be adjusted or abated.

II-1

3. OSHA Form No. 102 "Summary of Occupational
Injuries and Illnesses" must be posted by
February 1, each year and remain posted until
March 1.   Notices must be posted at all
establishments (except as stated in NOTE,
paragraph 3 below). This form is available
from the nearest OSHA Regional Office.
Penalty - Failure to post Form 102 as required
subjects employers to a proposed penalty of
$100, which may not be abated or adjusted.

4. An employer must report to the nearest OSHA
Area Office, orally or in writing, within 48
hours of any occupational accident fatal to one
or more employees, or which results in hospi-
talizations of five or more employees.
Penalty - Failure to give proper notice subjects an employer to a proposed penalty of
$200, which may not be adjusted or abated.

RECORD-
KEEPING

OSHA Regulation PART 1904 - "Recording and
Reporting Occupational Injuries and Illnesses"
prescribe the following recordkeeping:

1. OSHA Form 100 "Log of Occupational Injuries
and Illnesses". Must be retained for 5 years.

2. OSHA Form 101 "Supplementary Record of Oc-·
cupational Injuries and Illnesses". Must
be retained for 5 years.

3. OSHA Form 102 "Summary of Occupational
Injuries and Illnesses". (See paragraph 3 above:)

The Forms and instructions are available from
the nearest OSHA Regional or Area Office (pp. II-8-10).

Penalty - Failure to maintain OSHA Forms 100,
101 and 102 properly subjects an employer to
a proposed penalty of $100, which may not be
adjusted or abated.

In addition to the above records, certain employers shall receive, on a selective basis, OSHA
Form 103 "Occupational Injuries and Illnesses
Form". This Form is used in conjunction with
the OSHA Statistical program and must be completed and returned promptly upon receipt.

NOTE: As of October 4, 1972, OSHA revisions
to Part 1904 "Recording and Reporting Occupational Injuries and Illnesses" permit the
following procedures:

a. Employers must present or mail copies of
the Annual Summary (Form 102) to employees
who do not work at or report to any fixed
establishment regularly.

II-2

  b. Where operations have been closed down at some establishments during the year, it is not necessary to post summaries at those establishments.

  c. Construction industry employers may maintain records for job site operations subject to common supervision in an established central place. In such cases the name and address of the central place must be available at the job site, and personnel must be available at the central place during normal business hours to provide information from the records by telephone or mail.

  d. An employer with no more than 7 employees is required only to report fatalities or multiple hospitalization accidents. He need not prepare the Log (OSHA 100), Supplementary Record (OSHA 101) or Annual Summary (OSHA 102), except when such an employer has been notified that he has been selected to participate in an OSHA Statistical Survey.

INSPECTIONS 1. Inspections will be conducted by OSHA Compliance Officers without advance notice except in unusual circumstances. The inspection must be conducted during regular working hours.

  2. Compliance officers begin by interviewing the employer or his designated representative. Inspectors may interview anyone in private. Also, any employee may bring any conditions which he believes violates a standard to the attention of the inspecting officer.

  3. A representative of the employer and a representative authorized by employees shall be given an opportunity to accompany the Compliance Officer during the physical inspection. (The Department of Labor has ruled that the employer is not required to pay an employee's representative for the time spent on the accompanying tour of the premises). The Act does not require that there be an employee representative for each inspection tour.

  4. An inspection normally includes a check of a job site and examination of OSHA recordkeeping forms.

  5. Upon completion of an inspection, the Compliance Officer shall confer with the employer or his representative and advise him of any unsafe conditions or practices.

6. Follow-up inspections are mandatory for "imminent danger" situations or serious, willful and repeated violations. The Area Director has discretion as to follow-up inspections for non-serious violations.

TYPES OF VIOLATIONS

1. Serious Violations involve situations where a substantial probability exists of death or serious physical harm. Serious Violations may be assessed penalties up to $1,000.00 each.

2. Non-Serious Violations involve situations which may adversely affect safety or health of employees, but probably would not cause death or serious physical harm. Non-Serious Violations may be assessed penalties up to $1,000.00 each.

3. De Minimis Violations involve circumstances where failure to follow a standard has no immediate relationship to safety or health. No penalties are assessed for De Minimis violations.

4. Willful Violations involve circumstances where the employer violates OSHA intentionally or knowingly, or is aware that a hazardous condition existed and made no reasonably effort to eliminate the condition.
Repeated Violations are those where a second OSHA citation is issued for violation of the same condition as the first.
Willful and Repeated Violations may be assessed penalties of not more than $10,000.00.

CITATIONS

If a Compliance Officer observes what he considers violations of OSHA regulations during an inspection, he will issue a citation. The citations will:

1. Be issued in writing, on standardized OSHA forms.

2. List violation classifications as to Serious, Non-Serious, or De Minimis.

3. Cite specific regulations or parts of the Act violated.

4. Specify a date by which the violation must be abated. Serious violations may require immediate correction.

NOTE: Employers must post each citation at or near the location where the violation was found. (Penalty for failure to post citations may be up to $500.00.)

II-4

REDUCTION
OF
PENALTIES

In establishing proposed penalties for violations, a Compliance Officer establishes "unadjusted penalties" based on OSHA guidelines. He may then adjust the penalties downward based on the following:

1. "Good Faith" - A reduction of 20 percent may be given for good faith. Good faith is judged by two criteria--an awareness of the Act, exhibited by the effectiveness of the employer's Safety and Health program; and overt indications of the employer's desire to comply with the Act by specific displays of accomplishments.

2. Size - Up to 10 percent reduction may be given for size:
Less than 20 employees - 10% reduction
20 - 100 employees - 5% reduction
Over 100 employees - No reduction

3. History - A maximum reduction of up to 20 percent may be given for an employer's past history showing no violations or no repeated violations.

4. Abatement of condition - Penalties adjusted due to good faith, size and history may be further reduced by up to 50 percent, if the employer corrects the violation within the specified period shown in a citation. This reduction is made at the time the penalty is calculated to determine the final penalty to be assessed.

5. The total maximum reduction an organization can receive under 1,2,3,4 above is 75% of the initial assessment. For example, if a proposed penalty initially is $100, an employer qualifying for the 20% maximum allowed (each) for good faith, 20% for history, and 10% for size would have an "adjusted penalty" of $100 minus (50% x $100) = $50. The $50 "adjusted penalty" could then be reduced by 50% by correcting the violation within the specified period; the final total penalty assessed then would be $25.

CORRECTION
OF
VIOLATIONS

1. Citations that result from OSHA inspections specify for each alleged violation, a date by which it must be corrected. (See MCAA Safety Committee Bulletin Number 1 dated Feb. 23, 1972, for an OSHA citation in its entirety). The dates allowed for correction range from "Immediate" to about 30 days. OSHA

follow-up inspections may be made after the allowed abatement period to verify that the conditions cited have been corrected.

2. SECTION 17(d) OF THE ACT PROVIDES THAT ANY EMPLOYER WHO FAILS TO CORRECT A VIOLATION WITHIN THE PERIOD ALLOWED FOR ABATEMENT MAY BE ASSESSED A PROPOSED CIVIL PENALTY OF NOT MORE THAN $1000 FOR EACH DAY THAT THE VIOLA- TION CONTINUES FOLLOWING THE ALLOWED ABATE- MENT PERIOD.

3. If an employer contests an alleged violation in good faith, the period for abatement does not begin to run until the Review Commission affirms the citation. The number of days then allowed for abatement is the same as allowed originally. If an employer does not contest within 15 working days after receipt of the notice of proposed penalty, the citation and proposed assessment of penalties shall be deemed to be a final order of the OSHA Review Commission, and the original abatement date applies. It is important, therefore, for an employer receiving an OSHA citation to have a clear understanding of the required abatement dates and assure that violations are corrected within the allowable time; otherwise he is subject to costly fines for each day the violation continues after the deadline.

4. Informal Conference - An employer, employee, or employee representative may request an in- formal conference with the Regional Adminis- trator for the purpose of discussing any issues raised by an inspection, citation, notice of proposed penalty or notice of in- tention to contest (Paragraph 1903.19, PART 1903 "Inspections, Citations and Proposed Penalties"). Such a conference (or request for it) will not act as a stay for any 15- working day period for filing a notice of con- test as described in the following section. However, the conference provides an opportun- ity to discuss issues involved, and circum- stances related to any proposed citation. It is possible that the conference may result in settlement of issues concerned. (Note: This paragraph was added for the Second Printing of this handbook).

II-6

PROCEDURES FOR APPEALS

HOW TO
CONTEST
CITATIONS

OSHA citations that an employer considers unjust or unwarranted can be appealed, but the process is involved. The detailed procedures for appeals are contained in Title 29 CFR Part 2200, "Rules of Procedure". The following points apply:

1. Initially there are two things to be done by an employer who wishes to contest all or part of the citation he has received from OSHA, or the amount of any penalty proposed:

   a. Within fifteen working days from his receipt of the notification of proposed penalty, he must notify the Labor Department in writing of his intent to contest (this is called a Notice of Contest), and

   b. Upon receiving notice that the case has been docketed, he must notify his affected employees (and their union representative, if any) that he is contesting the case.

NOTICE OF
CONTEST

2. The Notice of Contest is a simple written statement by the employer, or his representative, that he intends to contest the OSHA citation. If the employer does not contest everything in the citation and proposed penalty, his letter should specify exactly what he is contesting.

3. FOR ANY CITATION OR ITEMS LISTED ON A CITATION WHICH THE EMPLOYER DOES NOT CONTEST, THE EMPLOYER MUST COMPLY WITH THE ABATEMENT PERIOD AND PAY TO THE DEPARTMENT OF LABOR WHATEVER MONETARY PENALTY IS SPECIFIED FOR SUCH CITATION OR ITEMS.

4. The Notice of Contest is sent to the Area Director office that issued the citation.

   NOTE: When Notice of Contest is filed in good faith and not solely for purposes of delay, the time specified for correcting the alleged violations does not begin to run until the final order of the Commission for those alleged violations contested. However, if only the amount of the penalty is contested, the original abatement period applies.

5. The OSHA Area Director forwards the employer's notice of contest to the Commission. The Commission's Executive Secretary then notifies the employer, the Secretary of Labor and all other parties to the case, of receipt of the Notice of Contest and the case docket number.

II-7

6. The Commission will furnish the employer with a copy of a notice for advising his affected employees.

NOTICE TO
EMPLOYEES

a. An employer then is required to give notice to all his affected employees as specified below:

   (1) For affected employees who are not represented by a union, the employer must:

       (a) Post a copy of the Notice of Contest at each place where the OSHA citation is required to be posted, and
       (b) Post a notice informing affected employees of their right to participate in the case (The Commission supplies forms for this).

   (2) If any affected employees are represented by a union for collective bargaining purposes, the employer is also required to:

       (a) Serve such union with a copy of the Notice of Contest and
       (b) Serve such union with a copy of the notice supplied by the Commission.

   (3) If more than one local union represents affected employees, each such union must be served.

   (4) If some affected employees are represented by a union and some are not, the employer must comply with all of the requirements listed in this section.

   (5) The employer must also notify the Commission that he has complied with these requirements for notifying affected employees (The Commission will furnish a return post card for this purpose).

COMPLAINT
AND
ANSWER

7. Within 20 days of the date on which OSHA receives the employer's notice of contest, counsel for the Secretary of Labor must file a written COMPLAINT with the Commission. A copy must be sent to the employer and all other parties to the case, if any. The Complaint includes the basis for the abatement period, and a justification of the proposed penalty.

II-8

The employer must file a written ANSWER to the
Complaint with the Commission within 15 days. A
copy must be sent to counsel for the Secretary
and to all other parties to the case, if any.
The Answer must be either a general denial of
the Complaint or respond specifically to those
statements in the Complaint. Any allegation
in the Complaint not denied in the Answer will
be deemed to have been admitted.

8. THE COMMISSION ENCOURAGES THE SETTLEMENT OF
CASES AND WILL GIVE HOSPITABLE CONSIDERATION TO
SETTLEMENTS THAT ARE AGREED UPON AT ANY STAGE
OF THE PROCEEDINGS.

HEARINGS

9. Hearings are adversary proceedings conducted
like trials in court. The hearing is usually
conducted in or near the community where the
alleged violation occurred.

After the Judge has heard the evidence and considered any written briefs, he will issue a
written decision. Fach party will receive a
copy. This decision becomes final 30 days after
its receipt by the Commission.

ABATEMENT
REQUEST

10. There are some instances where an employer has
not contested an OSHA citation (or availed himself of his right to appeal a Commission order)
but wishes to have a change in the abatement
date with which he is required to comply. If
the employer has made a good faith effort to
comply with that abatement period, but had not
been able to do so by the prescribed date because of factors beyond his control, he may file
a PETITION FOR MODIFICATION OF ABATEMENT. This
Petition is filed with the OSHA Area Director in
the same manner as a Notice of Contest and must
be filed no later than the end of the next working day following the date on which abatement
was to have been completed. It should state
why the abatement cannot be completed within the
prescribed time. Affected employees and their
union (if any) must also be notified.

OSHA is required to forward the Petition to the
Commission within 3 days of receipt.

Procedures are generally the same as with a Notice
of Contest, except that the employer must demon-
strate that he could not complete abatement for
reasons beyond his control.

The burden of proof is upon the employer.

REGIONAL OFFICES

of the

OCCUPATIONAL SAFETY AND HEALTH ADMINISTRATION

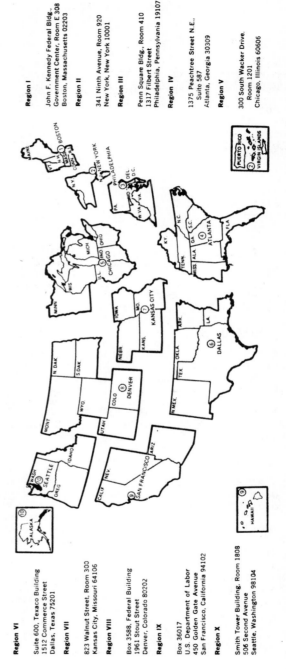

Region I

John F. Kennedy Federal Bldg.,
Government Center, Room E 308
Boston, Massachusetts 02203

Region II

341 Ninth Avenue, Room 920
New York, New York 10001

Region III

Penn Square Bldg., Room 410
1317 Filbert Street
Philadelphia, Pennsylvania 19107

Region IV

1375 Peachtree Street N.E.,
Suite 587
Atlanta, Georgia 30309

Region V

300 South Wacker Drive.
Room 1201
Chicago, Illinois 60606

Region VI

Suite 600, Texaco Building
1512 Commerce Street
Dallas, Texas 75201

Region VII

823 Walnut Street, Room 300
Kansas City, Missouri 64106

Region VIII

Box 3588, Federal Building
1961 Stout Street
Denver, Colorado 80202

Region IX

Box 36017
U.S. Department of Labor
450 Golden Gate Avenue
San Francisco, California 94102

Region X

Smith Tower Building, Room 1808
506 Second Avenue
Seattle, Washington 98104

OSHA AREA OFFICES

ALABAMA

Todd Mall
2047 Canyon Road
Birmingham, AL 35216

Room 801
Commerce Building
118 North Royal Street
Mobile, AL 36602

ALASKA

Room 214
Willholth Building
610 C Street
Anchorage, AK 99501

ARIZONA

Suite 910
Amerco Towers
2721 North Central Ave.
Phoenix, AZ 85004

CALIFORNIA

Room 514
Hartwell Building
19 Pine Avenue
Long Beach, CA 90802

Room 1706
100 McAllister Street
San Francisco, CA 94102

COLORADO

Squire Plaza Building
8527 West Colfax Ave.
Lakewood, CO 80215

CONNECTICUT

Room 617
Federal Building
450 Main Street
Hartford, CT 06103

FLORIDA

Room 204
Bridge Building
3200 E. Oakland Pk. Blvd.
Fort Lauderdale, FL 33308

U.S. Federal Office Bldg.
400 W. Bay St.
Box 35063
Jacksonville, FL 32202

GEORGIA

Room 723
1371 Peachtree St.,N.E.
Atlanta, GA 30309

Suite 201
Enterprise Bldg.
6605 Abercorn Street
Savannah, GA 31405

HAWAII

Suite 505
333 Queen Street
Honolulu, HI 96813

ILLINOIS

Room 1200
300 South Wacker Drive
Chicago, IL 60606

INDIANA

Room 423, 46 E. Ohio Street
Indianapolis, IN 46204

KENTUCKY

Room 561
600 Federal Place
Louisville, KY 40202

LOUISIANA

Room 1036
Federal Building
600 South Street
New Orleans, LA 70130

II-11

MARYLAND

Room 1110-A
Federal Building
31 Hopkins Plaza
Charles Center
Baltimore, MD 21201

MASSACHUSETTS

Custom   House Building
State Street
Boston, MA 02109

MICHIGAN

Room 626
Michigan Theatre Bldg.
220 Bagley Avenue
Detroit, MI 48226

MINNESOTA

Room 437
110 South Fourth St.
Minneapolis, MN 55401

MISSOURI

Room 1100
Davidson Building
1627 Main Street
Kansas City, MO 64103

Room 553
210 North 12th Street
St. Louis, MO 63101

MONTANA

Room 203
Central Park Bldg.
711 Central Avenue
Billings, MT 59102

NEBRASKA

Room 630
City National Bank Bldg.
16th and Harney Streets
Omaha, NE 68102

NEW HAMPSHIRE

Room 425
Federal Building
55 Pleasant Street
Concord, NH 03301

NEW JERSEY

Room 635
Federal Office Building
970 Broad Street
Newark, NJ 07102

NEW YORK

370 Old Country Road
Garden City
Long Island, NY 11530

Room 1405
90 Church Street
New York, NY 10007

Room 203
Midtown Plaza
700 East Water Street
Syracuse, NY 13210

NORTH CAROLINA

1361 East Morehead Street
Charlotte, NC 28204

OHIO

Room 5522
Federal Office Building
550 Main Street
Cincinnati, OH 45202

847 Federal Office Building
1240 East Ninth Street
Cleveland, OH 44199

Room 224
Bryson Building
700 Bryden Road
Columbus, OH 43015

Room 734
Federal Office Building
234 North Summit Street
Toledo, OH 43604

II-12

<u>*OKLAHOMA*</u>

Room 512
Petroleum Building
420 South Boulder Ave.
Tulsa, OK 74103

<u>*OREGON*</u>

Room 526
Pittock Block
921 S.W. Washington St.
Portland, OR 97205

<u>*PENNSYLVANIA*</u>

Suite 1010
1317 Filbert Street
Philadelphia, PA 19107

Room 445-D
Federal Building
1000 Liberty Avenue
Pittsburgh, PA 15222

<u>*PUERTO RICO*</u>

Room 328
Condominum San Alberto Bldg.
605 Condado Avenue
Santurce, PR 00907

<u>*RHODE ISLAND*</u>

Room 613
57 Eddy Street
Providence, RI 02903

<u>*TENNESSEE*</u>

Suite 302
1600 Hayes Street
Nashville, TN 37203

<u>*TEXAS*</u>

Room 6B1
Federal Building
1100 Commerce Street
Dallas, TX 75202

<u>*TEXAS*</u> *(continued)*

Room 325
U.S. Custom House Bldg.
Galveston, TX 77550

Room 802
Old Federal Office Bldg.
201 Fannin Street
Houston, TX 77002

Room 421
Federal Building
1205 Texas Avenue
Lubbock, TX 79401

<u>*UTAH*</u>

Suite 309
Executive Building
455 East Fourth South
Salt Lake City, UT 84111

<u>*VIRGINIA*</u>

Stanwick Building
Room 111
3661 Virginia Beach Blvd.
Norfolk, VA 23502

Room 8018
Federal Building
P.O. Box 10186
400 North Eighth Street
Richmond, VA 23240

<u>*WASHINGTON*</u>

1902 Smith Tower Bldg.
506 Second Avenue
Seattle, WA 98104

<u>*WISCONSIN*</u>

Room 906
Sheraton Schroeder Hotel
509 West Wisconsin Ave.
Milwaukee, WI 53203

REFERENCES AND ACKNOWLEDGEMENTS

1. Department of Labor

    "Safety and Health Regulations for Construction".
    (Title 29 - LABOR, Code of Federal Regulations,
    Chapter XVII, PART 1926, (formerly PART 1518))con-
    tained in Federal Register Volume 37, NUMBER 243,
    Saturday, December 16, 1972.

2. Department of Labor

    "Occupational Safety and Health Standards"
    (Title 29 - LABOR, Code of Federal Regula-
    tions, PART 1910) contained in Federal
    Register Volume 37, Number 202, October 18,
    1972.

3. Department of Labor, Occupational Safety and Health
    Administration

    "Compliance Operations Manual", January 1972, OSHA
    2006.

4. "Safety Manual", The Spohn Corporation, Cleveland, Ohio.

5. "Use and Care of Tools for Steamfitter-Pipefitter
    Journeymen and Apprentices" published by MCAA/UA
    National Joint Steamfitter-Pipefitter Apprenticeship
    Committee.

MCAA Safety Posters, some of which are shown in this
handbook, are available for order. See page 71,"MCAA
Mid-Year Membership Directory, 1972" for a listing of
posters and prices.

FEDERAL TAX LAW - CIRCULAR E

The Internal Revenue Service, Department of the Treasury, requires that every employer who witholds income tax from wages or who is liable for social security taxes must file a quarterly return. The *Employer's Tax Guide*, *Circular E* instructs the contractor on matters of deduction, filing, etc. The most important excerpts from Circular E are reprinted here for your study.

Department
of the
Treasury

**Internal
Revenue
Service**

1973

Circular E

Employer's
Tax Guide

Social Security Tax Base Increases

The maximum amount of wages subject to social security (FICA) taxes has been increased to $10,800 for wages paid in 1973.

Social Security Tax Rates Increase

The rates of FICA taxes have been increased to 5.85 percent each for both employee and employer for wages paid in 1973.

Federal Unemployment Tax Changes

The rate of tax has been increased to 3.28 percent of wages during 1973. Also, the rules for depositing unemployment tax have been changed. (See sections 14 and 17.)

The income tax withholding rates for 1973 have not changed from those used for 1972.

Publication 15

Important Notice

About Changes in Circular E

Because of the nature of changes in the tax law, we do not plan to revise Publication 15, Circular E, Employer's Tax Guide, for 1974.

The rates and tables contained in the 1973 issue of Circular E, for withholding income tax and social security (FICA) taxes from wages, remain in effect for 1974.

However, there have been two changes in the law that affect employment taxes. These are: (1) the maximum amount of wages subject to social security taxes has been increased to $12,600 for wages paid in 1974, and (2) the Federal unemployment tax rate has been decreased to 3.2 percent of the first $4,200 of wages paid to each employee each year after 1973. The rules for depositing Federal unemployment tax have been changed to reflect the decrease in the tax rate.

The instructions for Forms 940 and 941 will be revised accordingly.

Caution: *At the time this notice was printed, Congress was considering proposed legislation that could affect employment taxes. If these proposals become law the Internal Revenue Service will use news media to provide supplemental advice and instructions.*

Department of the Treasury
Internal Revenue Service

Notice 348 (11-73)

New Short Form 1040A for 1972

Your employees may be pleased to know that the Internal Revenue Service again has a Short Form 1040A. Employees who had just wages, not more than $200 of dividends, or not more than $200 of interest, and who do not plan to itemize their deductions may be able to use the new Short Form 1040A.

Please display the poster on page 47 of this circular on your bulletin board so that your employees can see how easy it is to fill out the new income tax return. With the information contained in the simplified instructions accompanying the new Short Form 1040A, most employees will be able to fill out their own returns. If assistance is needed, they should feel free to call or visit their nearest Internal Revenue Service office.

Calendar

On or Before January 31 and at End of Employment

Give each employee a completed Form W–2, Wage and Tax Statement. If Form W–2 is not required, give statement of social security wages and employee tax deducted. See section 21.

On or Before January 31

For Federal Unemployment Tax.—File Form 940, Employer's Annual Federal Unemployment Tax Return. If timely deposits of the full amount of tax due were made, ten additional days are allowed to file the return. See sections 2, 4, 14, and 17.

On or Before February 28

For Income Tax Withholding.—File Form W–3, Transmittal of Wage and Tax Statements, and include all required Internal Revenue Service copies (Copy A) of Form W–2 furnished employees for the preceding calendar year. See section 22.

On or Before April 30, July 31, October 31, and January 31

Deposit Federal unemployment tax due if more than $100.

File a quarterly return on Form 941 or 941E and pay balance of undeposited taxes. If timely deposits of the full amount of taxes due were made, ten additional days are allowed to file the return.

Before December 1

For Income Tax Withholding.—Request a new Form W–4 from each employee whose withholding allowances will be different in the next year from those shown on his current Form W–4.

On May 1

Discontinue the exemption from withholding for each employee who has not given you a new Form W–4E. See section 13.

Reminders

On Hiring New Employees

For Income Tax Withholding.—Ask each new employee for a Form W–4, withholding allowance certificate, or Form W–4E, exemption from withholding, as the case may be.

For Social Security (FICA) Taxes.—Record the number and name of each new employee from his social security card. If he does not have a number, have him apply for one on Form SS–5. See section 5.

On Each Payment of Wages or Annuities

For Income Tax Withholding.—Withhold tax from each wage payment or supplemental unemployment compensation benefit payment in accordance with the employee's Form W–4 and the applicable withholding rate. Withhold tax from annuity payments if requested on Form W–4P.

For Social Security (FICA) Taxes.—Withhold 5.85 percent from each wage payment. If employee reported tips, see section 7.

Table of Contents

Instructions

1. Purpose

This Circular provides employers with a summary of their responsibilities for the withholding, depositing, paying and reporting of Federal income tax, social security taxes, and Federal unemployment tax.

Except as shown in the tables on pages 13 through 16, these taxes apply to every employer who pays taxable wages to employees or who has employees who report tips. For more detailed information consult your local Internal Revenue office or see the Employment Tax Regulations.

The information in this circular applies to all employment taxes unless specific exception is made or where clearly inapplicable. For railroad retirement taxes, see instructions for Form CT–1.

2. Who Are Employers

An employer is any person or organization for whom an individual performs any service as an employee. It includes any person or organization paying wages to a former employee after termination of his employment.

For income tax withholding.—For purposes of income tax withholding, the term "employer" includes organizations that may be exempt from income tax, such as religious and government organizations, and that may or may not be exempt from social security and Federal unemployment taxes.

For Federal unemployment tax.—The term "employer" includes any person or organization that during 1972 or

2

1973: (a) paid wages of $1,500 or more in any calendar quarter, or (b) had one or more employees at any time in each of 20 calendar weeks.

3. Employer Identification Number

Show your employer identification number on all forms, attachments, and correspondence you send to Internal Revenue and the Social Security Administration.

You should have only one identification number. If you have more than one and have not been advised which one to use, notify the Internal Revenue Service Center where you file your return. State the numbers you have, the name and address to which each number was assigned, and the address of your principal place of business. The Service will tell you which number to use.

If you acquired the business of another employer, do not use the number assigned to him.

If you have not applied for an identification number, you should do so on Form SS-4, available from any Internal Revenue office or the Social Security Administration.

The employer identification number is also used to identify the tax accounts of various organizations and entities that have no employees.

4. Who Are Employees

Common Law Employees.—Everyone who performs services subject to the will and control of an employer both as to what shall be done and how it shall be done, is an employee for purposes of these taxes. It does not matter that the employer permits the employee considerable discretion and freedom of action, if the employer has the *legal right* to control both the method and the result of the services.

Though not always applicable, some of the characteristics of the term "employee" are that the employer has the right to discharge him and furnishes him with tools and a place to work.

In general, those in business for themselves are not employees. For example, physicians, lawyers, dentists, veterinarians, construction contractors, public stenographers, and others who follow an independent trade, business, or profession in which they offer their services to the public are not employees.

If the relationship of employer and employee exists, the description of the relationship by the parties as anything other than that of employer and employee is immaterial. It does not matter that the employee is designated as a partner, coadventurer, agent, or independent contractor; nor does it matter how the payments are measured, or how they are paid, or what they are called; nor whether the individual is employed full or part time.

No distinction is made between classes of employees. Superintendents, managers, and other supervisory personnel are employees. Generally, an officer of a corporation is an employee, but a director, in his capacity as such, is not. An officer who performs no services, or only minor ones, and who neither receives nor is entitled to receive remuneration is not considered an employee.

Whether the relationship of employer and employee exists under the usual common law rules will be determined in doubtful cases by an examination of the facts in each case.

If you wish a ruling as to whether a worker is an employee, file Form SS-8.

Statutory Employees.—For Federal unemployment tax, the term "employee" also means individuals described in paragraphs (a) and (d) below, who perform services for remuneration under the conditions in the last paragraph of this section.

For social security taxes, the term "employee" also means any of the following who perform services for remuneration under the conditions in the last paragraph of this section—

(a) an agent-driver or commission-driver engaged in distributing meat products, vegetable products, fruit products, bakery products, beverages (other than milk), or laundry or dry-cleaning services, for his principal;

(b) a full-time life insurance salesman;

(c) a homeworker working according to specifications furnished by the person for whom the services are performed, on materials or goods furnished by that person, and required to be returned to him or his designate; or

(d) a traveling or city salesman, other than an agent-driver or commission-driver. He must work full time soliciting and transmitting to his principal orders from wholesalers, retailers, contractors, operators of hotels, restaurants, or other similar establishments, for merchandise for resale or supplies for use in their business operations. He may, however, also engage in side-line activities for some other person.

Anyone within any of these categories is an employee for social security tax purposes and if in category (a) or (d), is an employee for Federal unemployment tax purposes, if his service contract contemplates he will personally perform substantially all the services. But an individual is not an employee if he has a substantial investment in facilities used to perform his services (other than facilities for transportation), or if the services are a single transaction not part of a continuing relationship with the person for whom they are performed.

5. Employee's Social Security Number

Record the name and number of each employee exactly as they appear on his social security card.

If a new employee does not have a social security card, have him obtain one at any Social Security office.

Federal Government Employers. Federal agencies should refer to Chapter 3000 of Part III of the Treasury Fiscal Requirements Manual. This chapter prescribes the procedures Federal agencies are to use in accounting for Federal income taxes withheld and social security taxes, all of which are collected in the U.S. Treasury at the time actually withheld.

State and Local Government Employers.—Generally, only the information in this circular which relates to withholding of Federal income tax is applicable to State and local government employment. *State and local government employers should not include social security contributions with their deposits of withheld income taxes. They should report withheld income taxes to Internal Revenue on Form 941E.* Information on reporting and acquiring social security coverage may be obtained from the appropriate State official.

Other Social Security Tax Information.—Employers of agricultural employees should refer to Circular A. Employers of household employees should refer to section 9 of this circular. For self-employment information refer to the instructions for Form 1040 and Schedule SE (Form 1040). The circulars and forms may be obtained from any Internal Revenue office.

3

6. What Are Taxable Wages

Wages subject to (a) income tax withholding (b) social security taxes, and (c) Federal unemployment tax consist of all remuneration in cash or other forms paid to an employee for services performed (for exceptions, see section 8 and pages 13 through 16). The word "wages" covers all types of employee remuneration, including salaries, vacation allowances, bonuses, and commissions. It is immaterial whether payments are based on the hour, day, week, month, or year, or on a piecework or percentage plan. For treatment of tips, see section 7. For reporting "Other Compensation" not subject to withholding on Form W–2, see section 21.

Wages paid in any form other than money are measured by the fair market value of the goods, lodging, meals, or other consideration given in payment for services. (See page 14.)

If wages are paid in a form other than money, make sure the tax is available for payment.

Amounts paid specifically for traveling or other ordinary and necessary expenses incurred or reasonably expected to be incurred in your business are not taxable wages. Traveling and other reimbursed expenses must be identified either by making a separate payment or by specifically indicating the separate amounts if both wages and expense allowances are combined in a single payment.

Supplemental Unemployment Compensation Benefits.—For purposes of withholding Federal income tax, supplemental unemployment compensation benefits are treated as if they were wages to the extent the benefits are includible in the gross income of the employee if they are paid to an employee because of his involuntary separation from employment whether or not temporary under a plan to which the employer is a party. The separation must be due directly to a reduction in force, discontinuance of a plant or operation or similar condition.

Moving Expenses.—Reimbursements made to an employee for moving expenses are not subject to withholding if, at the time paid, it is reasonable to believe that he is entitled to a deduction for them. You must withhold, however, if you believe that he is not entitled to a deduction. See Publication 521 (Tax Information on Moving Expenses) and sections 12 and 21.

7. What Are Taxable Tips

Cash tips received by an employee must be reported to you on or before the 10th day of the month following the month the tips were received. No report is required for any month in which the tips were less than $20. An employee must give you a Form 4070 reporting tips or a statement showing (a) his name, address and social security number, (b) your name and address, (c) the calendar month or period the statement is furnished for, and (d) the total amount of tips.

You must collect both employee social security tax and income tax on tips reported by the employee from wages due him or from other funds he makes available. The social security employer tax does not apply to tips. Discontinue collecting social security employee tax when the employee's combined wages and tips total $10,800 for the year. However, your liability for employer tax on wages continues until the wages other than tips total $10,800 for the year. Income tax withholding applies for the entire year on wages and tips, even though the social security limits have been reached.

Example: John Smith, a waiter employed by the Ritz Restaurant, receives regular wages of $120 a week. He also receives and reports to his employer tips of $100 a week. The employer must pay the employer's share of FICA tax on $6,240 for the year (52 weeks × $120 a week = $6,240). FICA deductions from wages of the employee will cease when his tips and wages combined reach $10,800 for the year, but income tax withholding continues.

Include in column 7 of Form 941 all tips reported by employees during the quarter until the tips and other wages reach $10,800 for the calendar year, whether or not sufficient employee funds were available to collect the tax. If you wish to report tips on a form separate from that used for wages, you may use Form 941a Continuation Sheet. List the tips in column 7 and show the total of this income in item 9, Schedule A, Form 941.

If, by the 10th day of the month following the month in which you received an employee's report on tips, you have not had sufficient employee funds available to permit deduction of the employee tax, you are no longer liable for collection of the tax. You must show the amount of the uncollected FICA tax on Form W–2. See instructions on Form 941.

Railroad employers should refer to the instructions for Form CT–1.

See the chart on page 16 for treatment of tips for Federal unemployment tax purposes.

8. Sick Pay

In reporting wages for Federal income tax purposes, employees may exclude certain payments made by an employer under a wage continuation plan for periods they were absent from work because of personal injuries or sickness.

If both the employee and the employer contribute to the plan, any benefits attributable to the employee's contributions are excludable without limit, but there are certain limitations on the exclusion of the benefits attributable to the employer's contributions.

To figure an employee's sick pay exclusion, you must first determine whether the sick pay is more than 75 percent of his weekly wage rate.

(1) If over 75 percent—The employee must be absent from work due to illness or injury for 30 calendar days before he qualifies for the exclusion. The exclusion thereafter may not exceed $100 a week.

(2) If 75 percent or less—The employee must be absent for seven calendar days before he qualifies for the exclusion, but the exclusion is limited to an amount computed at a regular weekly rate not to exceed $75. The waiting period applies to absences due to both illness and injury unless the employee is hospitalized at least one day during his absence. After 30 calendar days, the allowable weekly exclusion is increased to an amount not to exceed $100.

You are not required to withhold income tax from excludable sick pay provided your records show the amount of each payment, the excludable portion of it, and any other information you reasonably believe establishes the employee's entitlement to the exclusion.

All such payments must be included in the box "Wages paid subject to withholding" on Form W–2. However, if you maintain the records specified above, you may also show excludable sick pay on Form W–2 as a separate item, regardless of whether income tax was withheld from such amounts.

See page 15 for social security treatment of sick pay.

9. Household Employees

An employer may withhold income tax from remuneration paid to a household employee if both parties voluntarily agree.

A special test is provided for determining whether social security taxes apply to household services performed in or about your private home (other than on a farm operated for a profit). The taxes apply to all cash wages paid (regardless of when earned) to a household employee in a calendar quarter for household services, if you pay the employee $50 or more cash wages in the quarter. The $50-a-quarter test applies separately to each household employee.

Social security taxes do not apply to cash wages for domestic service in your home if performed by your spouse, or by your son or daughter under the age of 21.

Nor do these taxes apply to cash wages for domestic service performed by your mother or father unless (a) you have in your home a son or daughter who is under age 18 or has a physical or mental condition that requires the personal care of an adult for at least four continuous weeks in the quarter, and (b) you are a widow (widower) or are divorced, or you have a spouse in your home who, because of a physical or mental condition, is incapable of caring for your son or daughter for at least four continuous weeks in the quarter.

The taxes apply only to cash wages paid to household employees who meet the $50-a-quarter test. Checks, money orders, etc., are the same as cash, but the value of food, lodging, clothing, car tokens, and other noncash items furnished to a household employee is not subject to social security tax.

If you file Form 941 for business employees, you may include household employees on this form. Otherwise, report them on Form 942.

10. Partially Exempt Employment

If half or more of an employee's time in your employ in a pay period is spent performing services subject to the taxes, all amounts you paid to him for services performed in that pay period are taxable. If less than half the time is spent performing services subject to the taxes, none of the amount you paid to him is taxable for purposes of the Federal employment taxes including income tax withholding.

11. Payroll Period

For income tax withholding the payroll period is the period of service for which you ordinarily pay wages to an employee.

If you have a regular payroll period, withhold the income tax on the basis of that period even though the employee does not work the full period.

If you have no payroll period, withhold the tax as if the wages were paid on a daily or miscellaneous payroll period. This method requires a determination of the number of days (including Sundays and holidays) in the period covered by the wage payment. If the wages are unrelated to a specific length of time (for example, commissions paid on completion of a sale),

the number of days must be counted from the date of payment back to the latest of (a) the last payment of wages made during the same calendar year, (b) the date employment commenced if during the same calendar year, or (c) January 1 of the same year.

If an employee is paid for a period of less than one week and signs a statement under penalties of perjury that he is not working for wages subject to withholding for any other employer during the same calendar week, you may compute his withholding on the basis of a weekly, instead of a daily or miscellaneous, payroll period. If he later begins work for wages subject to withholding for another employer, he must notify you within 10 days. After that, you must compute his withholding on the basis of the daily or miscellaneous period.

12. Supplemental Wage Payments Including Tips

If you pay supplemental wages—such as bonuses, commissions, overtime pay, back pay including retroactive wage increases, or reimbursements for nondeductible moving expenses (as noted in section 6)—in the same payment with regular wages, withhold the income tax as if the total of the supplemental and regular wages were a single wage payment for the regular payroll period.

If you pay supplemental wages in a different payment from regular wages, the method of withholding income tax depends in part on whether you withheld income tax from the employee's regular wages.

If you have withheld income tax from the employee's regular wages, you may choose either of two methods for withholding income tax on the supplemental wages. (1) You may withhold at a flat percentage rate of 20 percent without allowing for any withholding allowances the employee claimed on his Form W-4, or (2) you may add the supplemental wages to the regular wages paid the employee within the same calendar year for the current or last preceding payroll period, determine the income tax to be withheld as if the aggregate amount were a single payment, subtract the tax already withheld from the regular wage payment, and withhold the remaining tax from the supplemental wage payment.

If you have not withheld income tax from the regular wages (as, for example, where an employee's withholding allowances exceed his wages) you must use method (2) above. You must add the supplemental wages to the regular wages you paid within the same calen-

dar year for the current or last preceding payroll period and withhold income tax on the total as though the supplemental wages and regular wages were one payment for a regular payroll period.

Tips Treated as Supplemental Wages.—If an employee receives regular wages and reports tips, determine the income tax to be withheld on the tips as if the amount of tips reported were a supplemental wage payment. If you have not withheld income tax from the employee's regular wages, you must add the reported tips, as you would other supplemental wages, to the regular wages you paid within the same calendar year for the current or last preceding payroll period and withhold income tax as though the tips and regular wages were one payment. If, however, you have withheld income tax from the regular wages, you may choose to withhold on the tips by either method (1) or (2) described above.

You must withhold income tax on the tips from wages (excluding tips) or from other funds the employee makes available.

Vacation Pay.—If an employee receives vacation pay for a vacation absence, the vacation pay is subject to withholding as though it were a regular wage payment made for the payroll periods during the vacation. If vacation pay is paid in addition to regular wages for the vacation period, the vacation pay is treated as a supplemental wage payment. If the vacation pay is for a time that exceeds the employer's normal payroll period, it should be allocated over the payroll periods for which it is paid.

13. Marital Status and Withholding Allowances

To determine income tax withholding, you must take these factors into account:

1. **Wages paid**—(Including tips reported) during the payroll period.

2. **Marital status.**—There are separate withholding tables for single and married employees.

3. **Withholding allowances.**—These correspond with the exemptions that will be allowable on the employee's Federal income tax return for himself, wife (husband), age, blindness, and dependents.

Each person claimed as a dependent must meet all of the following tests:

(a) Income.—Received less than $750 income. (If your child* was under 19 or was a full-time student, ignore this test.)

5

(b) *Support.*—Received over half his support from the employee.

(c) *Married dependents.*—Did not file a joint return with husband (wife).

(d) *Citizenship or residence.*—Was one of the following: a citizen or resident of the U.S., a resident of Canada, Mexico, the Republic of Panama or Canal Zone, or was an alien child adopted by and living with a U.S. citizen in a foreign country.

(e) *Relationship.*—Met one of the following tests:

(1) Was related to the employee in one of the following ways (if filing jointly, the dependent can be related to either husband or wife)—

*Child	Stepbrother	Son-in-law
Mother	Stepsister	Daughter-in-law
Father	Stepmother	The following if
Grand-	Stepfather	related by blood.
parent	Mother-in-law	Uncle
Brother	Father-in-law	Aunt
Sister	Brother-in-law	Nephew
Grandchild	Sister-in-law	Niece

* Child includes:
Your son, daughter, stepson, stepdaughter.
A child who lived in your home as a member of your family if placed with you by an authorized placement agency for legal adoption.
A foster child who lived in your home as a member of your family for the whole year.

(2) Was a person who lived in your home as a member of your family for the whole year.

4. Special withholding allowance.—This allowance is used only for withholding purposes. Each single person and each married person whose spouse is not also employed is entitled to one special withholding allowance. This allowance may not be claimed by either husband or wife when both are employed or by any employee who has two or more concurrent jobs.

5. Additional withholding allowances based on itemized deductions.—These are used only for withholding purposes and are allowable only if an employee claims them on Form W–4. For further details, see that form.

6. Exemption from withholding for nontaxable individuals.—If an employee gives you a completed Form W–4E certifying that he had no income tax liability for 1972 and expects none for 1973, his wages will be exempt from Federal income tax withholding. Employees must file Form W–4E annually if they wish to continue the exemption.

Form W–4, Employee's Withholding Allowance Certificate.—Ask each new employee to give you a signed Form W–4 on or before commencement of employment. A Form W–4 filed by a new employee is to be made effective upon the first payment of wages. If an em-

ployee fails to furnish a Form W–4, you must withhold tax as if he were a single person who has no withholding allowances.

Form W–4 will remain in effect until a new one is furnished.

Special Rules on Marital Status.—An employee whose husband or wife died during the current taxable year may claim the "married status" for that year on Form W–4.

An employee whose husband or wife died during either of the two preceding taxable years may claim married status if the employee (1) maintains as his home a household which is the principal place of abode of his child or stepchild for whom he is entitled to a deduction for an exemption and (2) was entitled to file a joint return with the husband or wife in the year of death.

A nonresident alien or a person married to a nonresident alien is considered single for withholding tax purposes.

An employee who qualifies as a "head of household" is considered single for withholding tax purposes.

14. Computing Employment Taxes

Income Tax Withholding From Wages.—You may determine the amount of income tax to withhold by using either the wage-bracket tables on pages 20 through 39, or the percentage method on pages 17 through 19.

Both methods distinguish between unmarried persons (either single or head of household) and married persons and take into account the number of allowances claimed by the employee.

Publication 493 contains alternative formula tables for percentage withholding, alternative methods for withholding, and combined income tax and social security tax withholding tables.

In addition to the amounts required to be withheld, employees may have additional amounts withheld by entering the desired additional amount on Form W–4.

Income Tax Withholding From Annuities.—If an annuitant files Form W–4P, Annuitant's Request for Federal Income Tax Withholding, you must withhold the amount of income tax he specifies, provided it is not less than $5 a month and does not reduce the amount of any annuity payment below $10. After the end of the year, give each such annuitant a Form W–2P showing the gross amount of annuity payments and the amount of income tax withheld during the year.

Social Security Taxes.—For wages paid in 1973, the social security tax

rate is 5.85 percent each on employees and employers, or a total of 11.7 percent of the first $10,800 of wages paid each employee by an employer. (See section 7 for information on tips.) A convenient table which produces the same result as multiplying each wage payment by 5.85 percent is shown on pages 40 to 46, inclusive.

Federal Unemployment Tax.—The Federal unemployment tax for 1973 is 3.28 percent of wages paid (for 1972, the rate is 3.2 percent). The tax applies to the first $4,200 of wages paid during the calendar year to each employee. You may take a credit against your Federal unemployment tax (not in excess of 2.7 percent of taxable wages) for contributions you paid into State unemployment funds.

The Federal unemployment tax is imposed upon employers and must not be deducted from the wages of employees.

15. Quarterly Return of Income Tax Withheld and Social Security (FICA) Taxes

Every employer who withholds income tax from wages or who is liable for social security taxes must file a quarterly return on Form 941 (Form 941E in certain cases, see page 7) unless the only wages paid are for domestic service or agricultural labor.

The dates on which the returns and tax payments are due, are as follows:

Quarters	Quarter ending	Due date
Jan.–Feb.–Mar.	Mar. 31	Apr. 30
April–May–June	June 30	July 31
July–August–Sept.	Sept. 30	Oct. 31
Oct.–Nov.–Dec.	Dec. 31	Jan. 31

However, if you made timely deposits of the full amount of the taxes due for the quarter, the return may be filed on or before the tenth day of the second month following the quarter.

Where to File.—File Form 941 with the Internal Revenue Service Center for the region in which your legal residence, principal place of business or office or agency is located. Addresses are:

New Jersey, New York City and counties of Nassau, Rockland, Suffolk, and Westchester	Internal Revenue Service Center 1040 Waverly Avenue Holtsville, N.Y. 11799
New York (all other counties), Connecticut, Maine, Massachusetts, New Hampshire, Rhode Island, Vermont	Internal Revenue Service Center 310 Lowell Street Andover, Mass. 01812
District of Columbia, Delaware, Maryland, Pennsylvania	Internal Revenue Service Center 11601 Roosevelt Boulevard Philadelphia, Pa. 19155
Alabama, Florida, Georgia, Mississippi, South Carolina	Internal Revenue Service Center 4800 Buford Highway Chamblee, Georgia 30006
Michigan, Ohio	Internal Revenue Service Center Cincinnati, Ohio 45298
Arkansas, Kansas, Louisiana, New Mexico, Oklahoma, Texas	Internal Revenue Service Center 3651 S. Interregional Hwy. Austin, Texas 78740

Alaska, Arizona, Colorado, Idaho, Minnesota, Montana, Nebraska, Nevada, North Dakota, Oregon, South Dakota, Utah, Washington, Wyoming	Internal Revenue Service Center 1160 West 1200 South St. Ogden, Utah 84405
Illinois, Iowa, Missouri, Wisconsin	Internal Revenue Service Center 2306 E. Bannister Road Kansas City, Mo. 64170
California, Hawaii	Internal Revenue Service Center 5045 East Butler Avenue Fresno, California 93730
Indiana, Kentucky, North Carolina, Tennessee, Virginia, West Virginia	Internal Revenue Service Center 3131 Democrat Road Memphis, Tennessee 38110

Do not report more than one calendar quarter on one return.

Use the preaddressed form mailed to you. If you misplace the form, request a new one in sufficient time to file a timely return. If you use a blank form, show your name and identification number as they appeared on previous returns.

If you go out of business or cease to pay wages, you must file a "final return." See section 22.

If you temporarily cease to pay wages or are engaged in seasonal activities, you must file a return for each quarter even though you have no taxes to report.

For social security taxes of household employees only, you may round each wage payment to the nearest whole dollar for purposes of determining taxes and reporting wages on returns. If you do so, you must round every wage payment made to all household employees during the same quarter.

Form 941.—Form 941 (see page 11) is required to be filed quarterly by all employers subject to income tax withholding, social security taxes, or both, except that (1) State and local government employers and other organizations that report only withheld income tax should use Form 941E, and (2) taxes on wages of household employees *may* be reported on Form 942. Taxes on wages of agricultural employees *must* be reported on Form 943. See section 7 for completion of Form 941 in the case of tips.

If Form 941 includes both business and household employees, identify the household employees in Schedule A by either grouping them after the listing of the business employees, under a heading "Household," or by inserting the letter "H" at the right of column 7, opposite the name of the employee. Household employees in a private home on a farm operated for profit should be reported as agricultural employees on Form 943.

Form 941E.—Form 941E is prescribed for reporting income tax withheld by State and local government employers and tax-exempt organizations that do not report social security taxes to the Internal Revenue Service.

It is also prescribed for reporting income tax withheld by certain payers of annuities.

A payer is the person making the annuity payment unless he is acting solely as an agent for another, in which case the other person is the payer.

Example 1. B, a bank, makes annuity payments only as an agent for an employee's trust. The trust is deemed to be the payer. If the trust has no employees whose wages are subject to FICA taxes, the trust should file Form 941E.

Example 2. A, an insurance company, makes annuity payments under contracts purchased by individuals. A has employees whose wages are subject to FICA taxes. A must file Form 941 and combine the income tax withheld from annuities with income tax withheld from wages.

Form 941a.—This is a continuation sheet for listing more employees than can be shown on Schedule A, Form 941.

It is also used in addition to Schedule A, Form 941, by employers who conduct business in more than one city or engage in two or more types of business, and who have an agreement with the Social Security Administration to list their employees by establishment or place of employment under the Establishment Reporting Plan. For more information on this method of reporting, obtain Inst. OAA–5019 "The Establishment Reporting Plan—A Statement for the Use of the Employer," from your local Social Security Administration District Office.

Magnetic Tape Reporting.—You are encouraged to submit your reports on magnetic tape instead of paper reports on Schedule A, Form 941, in accordance with Technical Instructions Bulletin #3 available from the Social Security Administration, Baltimore, Maryland 21235.

You may file Form 941 and composite returns of Form 941 for more than one employer on magnetic tape. For further details, see Revenue Procedure 72–37, available from any Internal Revenue Service Center.

Penalties.—Both criminal and civil penalties are provided for willful failure to file returns and pay tax when due, and for willfully filing false or fraudulent returns.

16. Adjustments

Every return on which an adjustment for a preceding quarter is reported must have an attached statement in duplicate explaining the adjustment, designating the return period to which the mistake relates, and giving other information called for in the instructions on the return.

If you deduct no tax, or less than the correct amount of tax (other than tax on tips), from any wage payment, you may deduct the amount of the undercollection from later payments to the employee. However, you are liable for any underpayment. Reimbursement is a matter for settlement between you and the employee.

If in any quarter you deduct more than the correct amount of tax from any wage payment, you should repay the overcollection to the employee. Keep as part of your records a written receipt from the employee showing the date and amount of the repayment. Every overcollection not receipted for by the employee must be reported and paid with the return for the quarter in which the overcollection was made.

Specific instructions for correcting mistakes in reporting withheld income tax and FICA taxes are on Form 941.

Since tips count against the $10,800 annual limit of wages subject to the employee social security tax but not against the employer tax, it may be necessary to report wages in item 14, Form 941, for the employer tax, even though the limit has been reached for employee tax. In these cases no more tips should be reported for the employee in column 7, Schedule A. But, continue to report wages paid the employee in column 6 until the wages other than tips reach $10,800. Since the tax on item 14 is figured at the combined rate of 11.7 percent, you should deduct in item 17 an amount equal to 5.85 percent of any wages not subject to the employee tax.

17. Payment of Federal Unemployment Tax

On or before January 31, every employer (see section 2) must file an unemployment tax return, Form 940, and deposit or pay the tax in full.

If timely deposits of the full amount of tax due are made, ten additional days are allowed to file the return.

Preaddressed Forms 940 are mailed each year to employers who filed returns for the previous year. Others should request Form 940 from Internal Revenue.

Federal unemployment tax must be deposited with an authorized commercial bank or a Federal Reserve bank. A Federal Tax Deposit Form 508 must accompany each deposit.

For deposit purposes, Federal unemployment tax must be computed on a quarterly basis. Any amount due must be deposited on or before the last day of

7

the first month following the close of the quarter. (For those who do not qualify as an employer until the second or third quarter, deposit requirements do not begin until the end of the second or third quarters.)

To determine whether you must make a deposit for any of the first three quarters in 1973, compute the total tax by multiplying by .0058 that part of the first $4,200 of each employee's annual wages that was paid during the quarter.

If the amount subject to deposit (plus the amount subject to deposit for any prior quarter but not deposited) is more than $100, deposit it during the first month following the quarter. If $100 or less, you do not have to deposit it, but you must add it to the amount subject to deposit for the next quarter.

If the tax reportable on Form 940, less amounts deposited for the year, is more than $100, you must deposit the entire amount on or before January 31. If your tax for the year (less any deposits) is $100 or less, you may either deposit the tax or pay it with Form 940.

18. Depositing Income Tax Withheld and Social Security (FICA) Taxes

Note. If any date shown falls on a Saturday, Sunday, or legal holiday, substitute the next regular workday.

Generally, you must deposit income tax withheld and social security taxes with an authorized commercial bank or a Federal Reserve bank. A Federal Tax Deposit Form 501 must accompany each deposit. If you employ agricultural labor, Federal Tax Deposit Form 511 must accompany each deposit of the taxes on their wages.

The amount of taxes determines the frequency of the deposits. The following rules and examples show how often you must make deposits:

(1) If at the end of a quarter the total amount of undeposited taxes is less than $200, you are not required to make a deposit. You may either pay the taxes directly to Internal Revenue along with your quarterly Form 941 or 941E or make a deposit.

Example: At the end of the second quarter the total amount of undeposited taxes for the quarter is $170. Since this amount is less than $200 you do not have to make a deposit. You may either pay the entire amount directly to Internal Revenue along with your quarterly Form 941 or 941E or make a deposit.

(2) If at the end of a quarter the total amount of undeposited taxes is $200 or more, you must deposit the

entire amount on or before the last day of the next month. If $2,000 or more, see rule 4.

Example: During the second quarter your taxes for each month of the quarter are $75. You must deposit $225 on or before July 31.

(3) If at the end of any month (except the last month of a quarter) the cumulative amount of undeposited taxes for the quarter is $200 or more and less than $2,000, you must deposit the taxes within 15 days after the end of the month. (This does not apply if you made a deposit for a quarter-monthly period that occurred during the month under the $2,000 rule in (4) below.)

Example A: During the second quarter your taxes for each of the first two months of the quarter are $300. You must deposit $300 within 15 days after both April 30 and May 31.

Example B: During the second quarter your taxes for each of the first two months of the quarter are $150. You must deposit $300 within 15 days after May 31.

Example C: During the second quarter your taxes are $500 for each month. You must deposit $500 within 15 days after both April 30 and May 31 and $500 on or before July 31.

(4) If at the end of any quarter-monthly period the cumulative amount of undeposited taxes for the quarter is $2,000 or more, you must deposit the taxes within three banking days after the end of the quarter-monthly period. (Quarter-monthly periods end on the 7th, 15th, 22d, and last day of any month.) In determining banking days exclude any local banking holidays observed by authorized commercial banks as well as Saturdays, Sundays, and legal holidays. The deposit requirements are considered met if: (a) you deposit at least 90 percent of the actual tax liability for the deposit period, and (b) if the quarter-monthly period occurs in a month other than the third month of a quarter, you deposit any underpayment with your first deposit that is required to be made after the 15th day of the following month. Any underpayment that is $200 or more for a quarter-monthly period that occurs during the third month of the quarter must be deposited on or before the last day of the next month.

Example A: During April, your taxes for each quarter-monthly period are $3,000. You must deposit $3,000 within three banking days after April 7, 15, 22, and 30.

Example B: During the second quarter your taxes for each quarter-monthly period are $700. You must deposit

$2,100 within three banking days after April 22, May 15, June 7, and June 30.

Summary of Deposit Rules for Income Tax Withheld and Social Security Taxes

Deposit rule	Deposit due
1. If at the end of a quarter the total amount of undeposited taxes is less than $200,	No deposit required. Either pay balance directly to Internal Revenue with quarterly return or make a deposit.
2. If at the end of a quarter the total amount of undeposited taxes is $200 or more,	On or before last day of next month. If $2,000 or more, see rule 4.
3. If at the end of any month (except the last month of a quarter), cumulative amount of undeposited taxes for the quarter is $200 or more and less than $2,000,	Within 15 days after end of month. (For the first two months of the quarter no deposit is required if you previously made a deposit for a quarter-monthly period that occurred during the month under $2,000 rule in 4.)
4. If at the end of any quarter-monthly period cumulative amount of undeposited taxes for the quarter is $2,000 or more,	Within 3 banking days after end of quarter-monthly period.

19. Use of Government Depositaries

How To Deposit Taxes.—Fill in a preinscribed Federal Tax Deposit Form 501 or Form 508, depending upon the type of tax being deposited, in accordance with instructions.

Send each Federal tax deposit form and a single payment covering the amount of taxes to be deposited to any commercial bank qualified as a depositary for Federal taxes, or to a Federal Reserve bank. Make checks or money orders payable to the bank you make your tax deposit with. Contact your local bank or Federal Reserve bank for the names of authorized commercial bank depositaries.

The timeliness of deposits is determined by the date received in a commercial bank depositary or Federal Reserve bank. A deposit received by the bank after the due date will be considered timely if you establish it was mailed two or more days before the due date.

How To Obtain Federal Tax Deposit Forms.—Preinscribed Federal tax deposit forms will automatically be sent to you after you apply for an identification number. If you need additional forms, order them from the Internal Revenue Service Center where you file.

If your branch offices make tax deposits, obtain a supply of Federal tax deposit forms and distribute them to the branches so they will be able to make deposits when due.

Do not use preinscribed forms of another employer. If you have not received Federal tax deposit forms by the due date of a deposit, mail your payment to the Internal Revenue office where you file your return. Make it payable to the Internal Revenue Service and show on it your name, identification number, address, kind of tax and period covered.

Record of Deposit.—Before making a deposit, enter the amount of payment on the form and stub, and record the check or money order number and date. Keep the stub for your records. The deposit portion of this form will not be returned to you, but will be used to credit your tax account as identified by your Employer identification number.

How to Claim Credit for Overpayments.—If you deposited more than the correct amount of taxes for a quarter, you may elect to have the overpayment refunded or applied as a credit to your next return. Show the appropriate amount in item 22, Form 941. Any amount shown there as a credit should be entered in Schedule B of your next return.

Penalty.—A five percent penalty is provided for failure, without reasonable cause, to make required deposits when due.

State and local government employers should deposit withheld income tax only. Contributions payable under a Federal-State social security coverage agreement should not be deposited. Any social security contributions and earnings reports should be sent to appropriate State officials.

---

20. Separate Accounting for Failure to Pay Taxes Withheld

Any employer who fails to pay over employee tax or income tax withheld, or who fails to make deposits, payments, or to file tax returns, as required, may be required to deposit such taxes in a special trust account for the U.S. Government and file monthly tax returns. Severe penalties are provided for failure to make such deposits and payments.

21. Statements for Employees and Annuitants

Wages Subject to Income Tax Withholding.—You must give a Form W–2 to each employee from whom you withheld income tax or would have withheld if he had claimed no more than one withholding allowance.

Supplemental Unemployment Compensation Benefits Subject to Income Tax Withholding.—Give a Form W–2 to each payee as if wages had been paid.

Other Compensation.—You must use Form W–2 in reporting all other compensation (amounts includible in gross income but not subject to income tax withholding) paid to an employee in the course of your trade or business. This rule applies to such payments, cash or noncash, if they amount to $600 or more in a calendar year or if the combined total of the payments and the employee's wages (subject to income tax withholding) amounts to $600 or more. (However, compensation under item (b), below, must be reported regardless of amount.) For **1972,** "Other Compensation" must be shown in the block designated on Form W–2, and should include such items as—

(a) Traveling or other expense allowances, even though excluded from wages under section 6 of this circular, unless an adequate accounting is made to the employer. However, reimbursements for moving expenses not deductible by the employee, from which income tax has been withheld, are wages and must be reported as such. Report any other reimbursements for moving expenses as "Other Compensation."

When making a reimbursement or payment of moving expenses to an employee (whether reimbursement or payment is made to such employee, to a third party for the benefit of such employee, or services are furnished in-kind to such employee), complete and furnish him a Form 4782 for each move made by such employee for which reimbursement or payment is made.

(b) The cost of group-term life insurance purchased for an employee to the extent that it is includible in his gross income under section 79 of the Internal Revenue Code.

(c) The value of noncash prizes or awards to retail commission salesmen who ordinarily are paid solely on a cash commission basis, if income tax is not withheld.

For **1973,** "Other Compensation" must be added to the wages paid.

Uncollected Tax on Tips.—If during the year sufficient funds were not available to collect the employee social security tax on tips, enter the uncollected amounts in the designated space on Form W–2.

Social Security Statements.—If a Form W–2 is required for any employee under the rules stated in this section and his wages were also subject to social security (FICA) tax, the Form W–2 must show the FICA wages paid (including tips reported) and the FICA tax withheld.

If Form W–2 is not required, you must give the employee a statement. A Form W–2 is preferred, but you may use Form SS–14 or a statement containing the identifying information and social security wage and tax information which appears on either Form W–2 or Form SS–14.

Example: Your wife employed a maid in the calendar year 1972. She paid the maid $48.50 in the first quarter, $60.00 in each of the second and third quarters, and $49.25 in the fourth quarter. The amounts paid in the second and third quarters would have been reported on the tax paid either on Form 941 or Form 942. The Form W–2 or Form SS–14 would show wages of $120 as subject to social security tax and FICA tax deductions of $6.24.

When To Furnish Form W–2 or Social Security Statement.—Give wage and tax statements on Form W–2 (or social security statements) for a calendar year and any corrected statements made in the year to employees not later than January 31 of the following year. However, if an employee leaves your employ before the end of the calendar year and is not expected to return to work within the calendar year, the statement must be given to him not later than 30 days after his last wages are paid.

Correcting Form W–2.—If it is necessary to correct a Form W–2 after it has been given to an employee, a corrected statement must be issued to the employee. Corrected statements should be clearly marked "Corrected by Employer." If a Form W–2 is lost or destroyed, a substitute copy clearly marked "Reissued by Employer" should be given to the employee.

Combined Federal-State Form.—To help employers in states where both Federal and State taxes must be withheld, a six-part Form W–2 may be used. The form may also be used where city or other subdivision taxes are withheld.

Privately Printed Forms.—The dimensions of substitute paper Forms W–2, used either for filing returns with IRS or for furnishing statements to em-

9

ployees, may be expanded from the dimensions of the official forms to make space available for conveying additional information (such as annual deductions from pay for hospitalization, life insurance, U.S. savings bonds, union or professional dues, etc.) not otherwise reported on these forms. This additional space may permit the elimination of a separate annual earnings statement or other notice which many employers provide to employees, in addition to the annual wage and tax statements. This approach may be used by, and should be particularly beneficial to, employers filing returns on magnetic tape.

See the Revenue Procedure on specifications for private printing of wage and tax statements and use of substitutes for information on the acceptable dimensions of paper forms. The Procedure is available at any Internal Revenue Service Center.

When employees are provided with wage and tax statements on privately printed forms, the legend "This information is being furnished on Form W–2 to the Internal Revenue Service and appropriate State officials" should be printed on the statement. Such statements are acceptable by the Internal Revenue Service when attached to employees' annual tax returns where the instructions to Form 1040 or Form 1040A call for attachment of Form W–2—and employees should be so advised.

Annuities.—If you are a payer of annuities, you must give each annuitant a Form W–2P. In recognition of the difficulty many payers have experienced in computing the taxable portion of the annuity, for 1972 you are required only to report the amount of tax withheld (if any) and the gross amount of annuity payments for the year. However, payers who are able to compute the taxable portion of the annuity should also furnish this information as it greatly assists annuitants in the preparation of their tax returns. Similarly, in the case of disability retirees receiving annuity or pension payments prior to their normal retirement date (when the payments are treated as wages subject to the sick pay exclusion of Code Section 105(d)) payers are encouraged to show the amount of the exclusion for the year on the statements.

Form W–2P is not applicable to:

1. Annuities and pensions wholly tax-exempt under the law; for example, social security benefits and Veteran's Administration Payments,

2. Periodic or other distributions effected by a trustee (non-employee plan) that are of a nature reportable on Form 1041 and Schedule E (Form 1041).

When to Furnish Form W–2P.—Give Form W–2P for a calendar year and any corrected statements made in the year to payees not later than January 31 of the following year. If the pension or annuity payments are terminated during the year, give Form W–2P to the payee within 30 days after the final payment.

Penalties.—Both criminal and civil penalties are provided for willful failure to furnish, in the manner and at the time prescribed, a statement showing the information required, or for willfully furnishing a false or fraudulent statement.

22. Report of Income Tax Withheld

On or before February 28, or when filing your final return on Form 941 or Form 941E, if final wages are paid before the end of the year, send copies A of all Forms W–2 issued for the year and Form W–3, Transmittal of Wage and Tax Statements, to the Internal Revenue Service Center with which you file. Payers of annuities are required to send copy A of all Forms W–2P issued for the year with both Form W–3P and Form W–3.

If your total payroll consists of a number of separate units or establishments, you may assemble the forms accordingly and submit a separate list for each establishment. In such case, submit a summary list, the total of which should agree with the corresponding entry on Form W–3.

If the number of forms is large, you may forward them in packages of convenient size. Identify each package with your name, number them consecutively, and place Form W–3 in package number one. At the top of W–3 show the number of packages. Forms that do not show employees' or payees' identifying numbers may be forwarded in the same package(s) as those that have these numbers but they must be grouped separately. Forms and packages sent by

mail are required by postal regulations to be sent by first class mail.

Any employee copies of Form W–2 which, after reasonable effort, cannot be delivered to employees should be sent to Internal Revenue with Form 941 or Form 941E for the second quarter of 1973, or with your final return if filed earlier. If you issue Forms W–2 and W–2P from branch establishments, instead of transmitting the undeliverable ones with your return, you may have the branches send them to their Internal Revenue Service Center.

Magnetic Tape Reporting.—You are encouraged to use computer produced magnetic tape for providing the information on Forms W–2 and W–2P to the Service. Many employers and payers achieve sizeable savings by using tape in lieu of paper documents. Interested parties should see Revenue Procedures 71–18 (covering reporting of fourth quarter FICA and Forms W–2 on a single tape) and 71–20 (magnetic tape filing of Forms W–2). The Director of the Internal Revenue Service Center with whom you are filing this form can provide you with further information on magnetic tape reporting.

23. Records To Be Kept

You must keep all records pertaining to employment taxes available for inspection by Internal Revenue if the need should arise.

No particular form has been prescribed for such records, but they should include the amounts and dates of all wage, annuity, and pension payments and tips reported; the names, addresses, and occupations of employees or payees receiving such payments; the periods of their employment; the periods for which they were paid while absent due to sickness or personal injuries and the amount and weekly rate of such payments; their social security numbers; their income tax withholding allowance certificates; the employer's identification number; duplicate copies of returns filed; and the dates and amounts of deposits made.

The records should be kept for a period of at least four years after the date the taxes to which they relate become due, or the date the taxes are paid, whichever is later.

Form **941**
(Rev. Jan. 1973)
Department of the Treasury
Internal Revenue Service

Employer's Quarterly
Federal Tax Return

(1)

SCHEDULE A—Quarterly Report of Wages Taxable under the Federal Insurance Contributions Act—**FOR SOCIAL SECURITY**
IF WAGES WERE NOT TAXABLE UNDER THE FICA MAKE NO ENTRIES IN ITEMS 1 THROUGH 9 AND 14 THROUGH 18

1. (First quarter only) Number of employees (except household) employed in the pay period including March 12th ▶	4	2. Total pages of this return including this page and any pages of Form 941a ▶	1	3. Total number of employees listed ▶	4

Please report each employee's name and number exactly as shown on his Social Security card.

4. EMPLOYEE'S SOCIAL SECURITY NUMBER	5. NAME OF EMPLOYEE (Please type or print)	6. TAXABLE FICA WAGES Paid to Employee in Quarter (Before deductions) Dollars / Cents	7. TAXABLE TIPS REPORTED (See page 4) If amounts in this column are not tips check here ☐ Dollars / Cents
000 00 0000			
000-00-0000	LEWIS J. SMALLWOOD	700 00	175 00
131-00-1200	WILLIAM H. HENRY	1,650 00	98 00
112-11-3111	JOHN HENRY JONES	1,650 00	
100-00-2973	BRYAN SMITH	1,425 00	200 00

(2) (3) (4)

If you need more space for listing employees, use Schedule A continuation sheets, Form 941a.
Totals for this page—Wage total in column 6 and tip total in column 7 ▶ | 5,425 00 | 473 00

8. TOTAL WAGES TAXABLE UNDER FICA PAID DURING QUARTER.
(Total of column 6 on this page and continuation sheets.) Enter here and in Item 14 below . . $ | 5,425 00 ◀

9. TOTAL TAXABLE TIPS REPORTED UNDER FICA DURING QUARTER. (If no tips reported, write "None.")
(Total of column 7 on this page and continuation sheets.) Enter here and in Item 15 below . . . | $ | 473 00

(5)

Name (as distinguished from trade name)	Date quarter ended
JOHN J. JONES	(6)	MARCH 73

Employer's name, address, employer identification number, and calendar quarter. (If not correct, please change)

Trade name, if any ▶ JONES DRY GOODS & RESTAURANT — Employer Identification No. 42-0000000

Address and ZIP code — 234 MAIN STREET, ANYTOWN, IOWA 00000

Entries must be made both above and below this line; if address different from previous return check here ☐

Name (as distinguished from trade name)	Date quarter ended
JOHN J. JONES	MARCH 73

Trade name, if any ▶ JONES DRY GOODS & RESTAURANT — Employer Identification No. 42-0000000

Address and ZIP code ▶ 234 MAIN STREET, ANYTOWN, IOWA 00000

T		FP	
FF		I	
FD		TOT	

(7)

10. TOTAL WAGES AND TIPS SUBJECT TO WITHHOLDING PLUS OTHER COMPENSATION . . . . ▶		5,898	00
11. AMOUNT OF INCOME TAX WITHHELD FROM WAGES, TIPS, ANNUITIES, etc. (See instructions)	(9)	602	10
12. ADJUSTMENT FOR PRECEDING QUARTERS OF CALENDAR YEAR			
13. ADJUSTED TOTAL OF INCOME TAX WITHHELD		602	10
14. TAXABLE FICA WAGES PAID (Item 8) . . . . $ 5,425.00 multiplied by 11.7%=TAX		634	73
15. TAXABLE TIPS REPORTED (Item 9) . . . . $ 473.00 multiplied by 5.85%=TAX		27	67
16. TOTAL FICA TAXES (Item 14 plus Item 15) . . . . (10) ▶		662	40
17. ADJUSTMENT (See instructions)			
18. ADJUSTED TOTAL OF FICA TAXES . . . . ▶		662	40
19. TOTAL TAXES (Item 13 plus Item 18)		1,264	50
20. TOTAL DEPOSITS FOR QUARTER (INCLUDING FINAL DEPOSIT MADE FOR QUARTER) AND OVERPAYMENT FROM PREVIOUS QUARTER LIST IN SCHEDULE B (See instructions on page 4) . . . . ▶		1,264	50

(8) (11) (12) (13)

Note: If undeposited taxes at the end of the quarter are $200 or more, the full amount must be deposited with an authorized commercial bank or a Federal Reserve bank. This deposit must be entered in Schedule B and included in item 20.

21. UNDEPOSITED TAXES DUE (ITEM 19 LESS ITEM 20—THIS SHOULD BE LESS THAN $200). PAY TO INTERNAL REVENUE SERVICE AND ENTER HERE . . . . ▶	-0-	
22. IF ITEM 20 IS MORE THAN ITEM 19, ENTER EXCESS HERE ▶ $ AND CHECK IF YOU WANT IT ☐ APPLIED TO NEXT RETURN, OR ☐ REFUNDED.		

23. If not liable for returns in succeeding quarters write "FINAL" here ▶ _____ and enter date of final payment of taxable wages here ▶

Under penalties of perjury, I declare that I have examined this return, including accompanying schedules and statements, and to the best of my knowledge and belief it is true, correct and complete.

Date **April 29, 1973** Signature *John J. Jones* Title (Owner, etc.) **Owner**

(14)

Exhibit of completed Form 941—circled items are explained on next page.

11

(1) Be sure to answer items 1 (first quarter only), 2, and 3.

(2) Be sure to enter each employee's social security number correctly so that each employee's wages will be credited to his earnings record. If you don't know the number, follow the instructions in section 5 of this circular.

(3) Please type or print the full name of the employee as shown on his social security card.

(4) Show in column 6 the total amount of taxable wages you paid each employee during the quarter. In column 7, show the total amount of tips each employee reported during the quarter. See section 16 for yearly limitation on amounts to be shown in columns 6 and 7. If you do not have employees who had tips, column 7 may be used for any payroll or State unemployment information that will facilitate your bookkeeping. You must then enter a checkmark in the block in column 7.

(5) Enter at the bottom of column 6 the total wages, and at the bottom of column 7 the total tips on this page. In item 8, enter the total wages on this and any other pages. Enter in item 9 the total tips on this and any other pages. The amount in items 8 and 9 should equal the amounts in items 14 and 15, respectively.

(6) This space is for your name, address, employer identification number, and the last month in the quarter covered by the return. These entries are ordinarily made here before mailing the Form 941 to you. Be sure to write this information on your copy of the return. If it is incorrect, or if your address changes, draw a line through the error and type or print the necessary correction.

(7) Enter the combined amounts of the total wages paid, the tips reported and other compensation paid to your employees for this quarter. Include all compensation whether or not subject to income tax or FICA taxes. Do not include the amount of any annuity or supplemental unemployment compensation benefit from which income tax has been withheld.

(8) Enter the total income tax withheld from wages, tips, supplemental unemployment compensation benefits, and annuities for the quarter.

(9) Enter total FICA wages from item 8, Schedule A.

(10) Enter total tips reported by employees from item 9, Schedule A.

(11) Multiply the FICA wages by the rates applicable (11.7%) that is, 5.85% employer tax and 5.85% employee tax. The total in this case is $634.73.

(12) Multiply the tips reported by 5.85%. The total in this case is $27.67. See sections 7 and 16 of this circular if all of this tax on tips was not collected from your employees.

(13) No tax is due. The total tax on item 19 ($1,264.50) was deposited in full.

(14) Please remember to sign the return, date it, and show your title.

Nine out of 10 working people in the United States are now building protection for themselves and their families under the social security program. The three kinds of monthly benefits under Social Security are:

1. Retirement—at age 65. (Reduced benefits are payable as early as 62.)

2. Disability—when a worker under 65 becomes unable to work because of a disability.

3. Survivors—when a worker dies.

In addition to cash benefits, health insurance benefits are available for certain categories of people, whether or not such people are retired.

Special classes of employment and special types of payment	Treatment under different employment taxes		
	Income tax withholding	Social security	Federal unemployment
Agricultural labor	See Circular A	See Circular A	See Circular A
Aliens:			
a. Resident:			
1. Service performed in U.S.	Same as U.S. citizen	Same as U.S. citizen; service as crew member of foreign vessel or aircraft exempt if any part is performed outside U.S.	Same as U.S. citizen
2. Service performed outside U.S.	Same as U.S. citizen	Exempt unless on or in connection with an American vessel or aircraft and either performed under contract made in U.S. or alien is employed on such vessel or aircraft when it touches U.S. port.	
b. Nonresident working in U.S.:			
1. Canadians and Mexicans entering U.S. frequently in transportation service across boundary or in construction or operation of waterway, bridge, etc., at boundary.	Exempt under the conditions stated in the regulations.	Exempt if railroad service	Exempt if railroad service
2. Other Canadians and Mexicans entering U.S. frequently to work.	Same as U.S. citizen	Same as U.S. citizen	Same as U.S. citizen
3. Workers from any foreign country or its possession lawfully admitted on a temporary basis to perform agricultural labor.	Exempt when performing agricultural labor.	Exempt	Same as U.S. citizen
4. Student, scholar, trainee, teacher, etc., as nonimmigrant alien under section 101(a)(15) (F) or (J) of Immigration and Nationality Act.	Taxable unless excepted by regulations.	Exempt if service is performed for purpose specified in section 101 (a)(15) (F) or (J) of Immigration and Nationality Act.	
5. All other nonresidents working in U.S.	Taxable unless excepted by regulations.	Same as U.S. citizen; service as crew member of foreign vessel or aircraft exempt if any part performed outside U.S. and employer is not "American employer."	Same as U.S. citizen
c. Nonresident working on American vessel or aircraft outside U.S.	Exempt	Taxable if under contract made in U.S. or worker is employed on vessel or aircraft when it touches U.S. port.	
Deceased worker's wages paid to beneficiary or estate after the year of worker's death.	Exempt	Exempt after 1972.	Taxable
Disabled worker's wages:			
Wages paid after year in which worker became entitled to disability insurance benefits under the Social Security Act.	See section 8	Exempt after 1972, if worker did not perform any service for employer during period for which payment is made.	Exempt
Dismissal or severance pay	Taxable	Taxable	Taxable
Domestic service in college clubs, fraternities and sororities.	Exempt (Taxable if both employer and employee voluntarily agree.)	Exempt if paid to regular student; also if less than $50 is earned by employee in a quarter from an income tax-exempt employer.	Exempt
Employers whose taxability depends on number of employees.	Taxable if one or more employees.	Taxable if one or more employees.	Taxable if during 1972 or 1973 you: 1. paid wages of $1,500 or more in any calendar quarter, or 2. had one or more employees at any time in each of 20 calendar weeks.
Family employees:			
a. Son or daughter under 21 employed by parent (or by partnership consisting only of parents); wife employed by husband or husband employed by wife.	Taxable	Exempt	Exempt
b. Parent employed by a son or daughter.	Taxable	Taxable if in course of the child's business; exempt if not in the course of the child's business. For household work in private home of child refer to section 9.	Exempt

Special classes of employment and special types of payment	Treatment under different employment taxes		
	Income tax withholding	Social security	Federal unemployment
Federal employees:			
a. Members of uniformed services; Peace Corps volunteers.	Taxable	Taxable	Exempt
b. All others	Taxable	Exempt if under civil service retirement system and in certain other cases.	Exempt unless worker is a seaman performing services on or in connection with American vessel operated by general agent of Secretary of Commerce. (Coverage under Title XV of Social Security Act.)
Fishing and related activities, employment in connection with:			
a. Salmon or halibut	Taxable	Taxable	Taxable
b. Other fish, sponges, etc.	Taxable	Taxable	Exempt unless on vessel of more than 10 net tons.
Foreign Government and international organizations.	Exempt	Exempt	Exempt
Foreign service by U.S. citizens:			
a. As U.S. Government employee	Taxable	Same as within U.S.	Exempt (See also "Federal employees.")
b. For foreign subsidiaries of domestic corporations and other private employers.	Exempt under the conditions stated in the regulations.	Exempt unless (1) U.S. citizen works for American employer or (2) a domestic corporation by agreement covers U.S. citizens employed by its foreign subsidiaries.	Exempt unless (1) on American vessel or aircraft and work is performed under contract made in U.S. or worker is employed on vessel when it touches U.S. port or (2) beginning in 1972, U.S. citizen works for American employer (except in a contiguous country with which the U.S. has an agreement for unemployment compensation or in the Virgin Islands).
Homeworkers (industrial):			
a. Common law employees	Taxable	Taxable	Taxable
b. Statutory employees. (See sec. 4.)	Exempt	Taxable if paid $50 or more in cash in quarter.	Exempt
Household workers (domestic service in private homes; farmers, see Circular A).	Exempt (Taxable if both employer and employee voluntarily agree.)	Taxable if paid $50 or more in cash in quarter.	Exempt
Insurance agents or solicitors:			
a. Full-time life insurance salesmen	Taxable only if employees under common law.	Generally taxable, regardless of common law.	Exempt if not a common law empluyee or if paid solely by commissions.
b. Other salesmen of life, casualty, etc., insurance.	Taxable only if employees under common law.	Taxable only if employees under common law.	Exempt if not a common law employee or if paid solely by commissions.
Interns working in hospitals	Taxable	Taxable if hospital is subject to social security.	Exempt
Meals and lodging (For household employees, agricultural labor, and service not in the course of the employer's trade or business, see "Noncash payments" below.)	(a) Meals—taxable unless furnished for employer's convenience and on his premises. (b) Lodging—taxable unless furnished on employer's premises, for his convenience, and as condition of employment.	Taxable Taxable	Taxable Taxable
Members of religious orders who have taken a vow of poverty performing duties required by the order.	Exempt (Taxable if both employer and employee voluntarily agree.)	Exempt unless employer, i.e. religious order or autonomous subdivision thereof, makes irrevocable election of coverage for entire active membership and for its lay employees.	Exempt
Ministers of churches performing duties as such, and members of religious orders who have not taken a vow of poverty, performing duties required by the order.	Exempt (Taxable if both employer and employee voluntarily agree.)	Exempt	Exempt

14

Special classes of employment and special types of payment	Treatment under different employment taxes		
	Income tax withholding	Social security	Federal unemployment
Moving expenses:			
a. Reimbursement for moving expenses you believe are deductible by employee.	Exempt	Exempt	Exempt
b. Reimbursement for moving expenses you believe are not deductible by employee.	Taxable	Taxable	Taxable
Newspaper carrier under 18 delivering to consumers.	Exempt (Taxable if both employer and employee voluntarily agree.)	Exempt	Exempt
Newspaper and magazine vendors buying at fixed prices and retaining excess from sales to consumers.	Exempt (Taxable if both employer and employee voluntarily agree.)	Exempt	Exempt
Noncash payments:			
a. For household work, agricultural labor, and service not in the course of the employer's trade or business.	Exempt (Taxable if both employer and employee voluntarily agree.)	Exempt	Exempt
b. To certain retail commission salesmen ordinarily paid solely on a cash commission basis.	Optional with employer	Taxable	Taxable
Nonprofit organizations:			
a. Religious, educational, charitable, etc., organizations described in sec. 501(c)(3) exempt from income tax under sec. 501(a), I.R.C.	Taxable	Exempt unless employee (1) concurs in employer's certificate effecting coverage (Form SS–15) or is hired or rehired after the quarter in which the certificate is filed, and (2) earns $50 or more in quarter.	Exempt
b. Other organizations exempt under sec. 501(a) (other than a pension, profit-sharing, or stock bonus plan described in sec. 401(a)) or under sec. 521, I.R.C.	Taxable	Taxable if $50 or more is earned by employee in quarter.	
Patients employed by hospitals	Taxable	Taxable if hospital is subject to social security.	Exempt
Railroads, etc. Payments subject to Railroad Retirement Tax Act and Railroad Unemployment Insurance Act.	Taxable	Exempt	Exempt
Retirement and pension payments	Exempt if payments are taxable to employee as annuities or excludable from gross income under section 104(a)(4), I.R.C. unless payee files Form W–4P.	Exempt	Exempt
Salesmen:			
a. Common law employees	Taxable	Taxable	Taxable
b. Statutory employees (referred to in paragraphs (a) and (d) in section 4).	Exempt	Taxable	Taxable
Service not in the course of the employer's trade or business, other than on a farm operated for profit (cash payments only).	Taxable only if $50 or more is earned in a quarter and employee works on 24 or more different days in that quarter or in the preceding quarter.	Taxable if paid $50 or more in cash in quarter.	Taxable only if $50 or more is earned in a quarter and employee works on 24 or more different days in that quarter or in the preceding quarter.
Sickness or injury payments under:			
a. Workmen's compensation law or contract of insurance.	Exempt	Exempt	Exempt
b. Certain employer plans	See section 8	Exempt	Exempt
c. No employer plan	See section 8	Exempt after end of 6 calendar months after calendar month employee last worked for employer.	Exempt after end of 6 calendar months after calendar month employee last worked for employer.
Standby employee (man 65 or over, woman 62 or over) doing no actual work in period for which paid.	Taxable	Exempt	Exempt, but both must be age 65 or over.

15

Special classes of employment and special types of payment	Treatment under different employment taxes		
	Income tax withholding	Social security	Federal unemployment
State governments and political subdivisions, employees of:		Exempt (except certain transportation services). Coverage under Social Security Act may be obtained only by agreement between State and Secretary of Health, Education, and Welfare.	
a. Fees of public officials	Exempt		Exempt
b. Salaries and wages	Taxable		
Students:			
a. Student working for private school, college or university, if enrolled and regularly attending classes.	Taxable	Exempt	Exempt
b. Student performing services for auxiliary nonprofit organization which is organized and operated exclusively for the benefit of and supervised or controlled by a private educational institution at which the student is enrolled and regularly attending classes.	Taxable	Exempt after 1972	Exempt
c. Student working for public school, college or university, if enrolled and regularly attending classes and student nurse working for public hospital.	Taxable	See state governments and their political subdivisions, above.	Exempt
d. Spouse of student, if such spouse is advised at the time service commences that (1) the employment is provided under a program to provide financial assistance to the student by the school, college or university, and (2) the employment will not be covered by any program of unemployment insurance.	Taxable	See nonprofit organizations and state governments and their political subdivisions, above.	Exempt
e. Student under age 22 enrolled in a full time program at a nonprofit or public educational institution. Institution must normally maintain a regular faculty and curriculum and normally have a regularly organized body of students where its educational activities are carried on. Student's service must be taken for credit at such institution. It must combine academic instruction with work experience. It must be an integral part of such program, and the institution must have so certified to the employer.	Taxable	Taxable	Exempt unless program established for or on behalf of an employer or group of employers.
f. Student nurse working for hospital.	Taxable	Exempt	Exempt
Supplemental unemployment compensation plan benefits:	Taxable	Exempt	Exempt
Tips if $20 or more in a month	Taxable	Taxable for employee only	Taxable to the extent reported in writing to employer and considered in determining employee's compensation under minimum wage law.
Tips if less than $20 in a month	Exempt from withholding. Taxable to employee.	Exempt	Exempt
Wage limit: Maximum of taxable wages paid each employee by same employer in same calendar year.	Unlimited	$10,800. Employee tips may not be used in computing maximum for purposes of employer tax. (Limit for new owner of business is reduced by predecessor's wage payments in certain cases.)	$4,200.
Workmen's compensation	Exempt	Exempt	Exempt

Income Tax Withholding—Percentage Method

(This section may be disregarded by any employer using the wage-bracket tables on pages 20 through 39)

Employers who prefer not to use the wage-bracket tables in computing the amount of income tax to be deducted and withheld from a payment of wages to an employee will make a percentage computation based upon the following percentage method withholding table and the appropriate rate table.

Percentage method income tax withholding table.

Payroll period	Amount of one withholding allowance
Weekly	$14.40
Biweekly	28.80
Semimonthly	31.30
Monthly	62.50
Quarterly	187.50
Semiannual	375.00
Annual	750.00
Daily or miscellaneous (per day of such period)	2.10

The steps in computing the income tax to be withheld under the percentage method are as follows:

(1) Multiply the amount of one withholding allowance (see table above) by the number of allowances claimed by the employee;

(2) Subtract that amount from the employee's wages;

(3) Determine amount to be withheld from appropriate table on pages 18 and 19.

Example.—An unmarried employee has a weekly payroll period, for which he is paid $95, and has in effect a Form W-4 claiming one personal allowance and the special withholding allowance. His employer, using the percentage method, computes the income tax to be withheld as follows:

(1) Total wage payment	$95.00
(2) Amount of one allowance	$14.40
(3) Number of allowances claimed on Form W-4 (including the special withholding allowance)	2
(4) Line 2 multiplied by line 3	28.80
(5) Amount subject to withholding (line 1 minus line 4)	66.20
(6) Tax to be withheld on $66.20 from Table 1—Single person, page 18:	
Tax on first $35.00	3.36
Tax on remainder $31.20 @ 18%	5.62
Total to be withheld	$8.98

In determining the amount of income tax to be deducted and withheld, an employer may reduce the last digit of the wage amount to zero, or may compute the wage amount to the nearest dollar. For example, if the weekly wage is $87.43, he may eliminate the last digit and determine the income tax on the basis of a wage payment of $87.40, or he may determine the tax on the basis of a wage payment of $87.

Annualized Income Tax Withholding.—Employers may determine the amount of income tax to be withheld on annualized wages under the Percentage Method of Withholding for an annual payroll period; and then prorate the tax back to the payroll period.

Example.—A married person claiming three personal allowances and the special withholding allowance is paid $110 a week. An employer will multiply the weekly wages of $110 by 52 weeks to determine an annual wage of $5,720. He would subtract $3,000 and arrive at a balance of $2,720. The table for the annual payroll period indicates that the tax to be withheld for $2,720 is $210, plus 16% of the excess over $2,050, or a total of $317.20. The annual tax of $317.20, when divided by 52 to arrive at the weekly share, equals $6.10.

Adjusting Wage Bracket Withholding for Employees Claiming More Than 10 Allowances

The wage bracket tables are extended only to 10 allowances. Allowances in excess of 10 will occur more often now because of the "special withholding allowance", additional allowances for itemized deductions, and the structure of the tax system which usually makes it advisable for employees to claim all the withholding allowances to which they are entitled.

Employers using the percentage method have no problem because that method is adaptable to any number of allowances. However, it is not feasible to extend the wage bracket tables beyond 10 allowances because the tables would become too bulky for convenient use. Therefore, an alternative which enables users of the wage bracket tables to adapt them to employees who have more than 10 allowances follows:

(1) Multiply the number of allowances in excess of 10 by the value for the payroll period. The allowance values are:

Payroll period	Allowance value
Weekly	$14.40
Biweekly	28.80
Semimonthly	31.30
Monthly	62.50
Daily or miscellaneous	2.10

(2) Subtract the result from the employee's wages.

(3) Find and withhold the tax in the column for 10 allowances for the wages net of the subtracted amount.

For example, in the case of a married employee with weekly wages of $583 who claims 12 withholding allowances on Form W-4, his employer using the wage bracket tables would determine the weekly amount to be withheld as follows:

(1) Multiply 2 allowances (the excess over 10) by $14.40 (the allowance value for a weekly payroll period) = $28.80.

(2) Subtract $28.80 from the weekly wages of $583 = $554.20.

(3) The net amount, $554.20, falls in the weekly wage bracket "at least" $550 "but less than" $560. The employer then finds the tax amount, $84.90, in the column for 10 allowances for the $550—$560 wage bracket and withholds this amount from the employee's weekly wages.

This method is voluntary on the part of employers. Employers using the wage bracket tables may continue to deduct and withhold the amount in the "10 or more" column when the employee has more than 10 withholding allowances. Or, they may use any of the other methods described in this circular or the Supplement to Circular E (Publication 493).

Other Methods of Income Tax Withholding

A. Publication 493, available from any Internal Revenue office, contains:

(1) Alternative formula tables for percentage method withholding,

(2) Alternative methods of withholding income tax, and

(3) Combined income tax and social security tax withholding tables.

B. Employers may develop their own methods of determining the amount of tax to be deducted and withheld. An employer's own method, however, must result in substantially the same amount of withholding as the percentage method (shown above) for the payroll period.

The following table shows the maximum amount of deviation there can be (on an annual basis) between an employer's own withholding method and the annual percentage rate method:

If the tax required to be withheld under the annual percentage rate schedule is—	The maximum permissible annual deviation is—
$0 to $10	$10
$10 to $100	$10, plus 10% of excess over $10
$100 to $1,000	$19, plus 3% of excess over $100
$1,000 or over	$46, plus 1% of excess over $1,000

17

Tables for Percentage Method of Withholding

TABLE 1. WEEKLY Payroll Period

(a) SINGLE person—including head of household:

If the amount of wages is:		The amount of income tax to be withheld shall be:	
Not over $11		0	
Over—	But not over—		of excess over—
$11	—$35	14%	—$11
$35	—$73	$3.36 plus 18%	—$35
$73	—$202	$10.20 plus 21%	—$73
$202	—$231	$37.29 plus 23%	—$202
$231	—$269	$43.96 plus 27%	—$231
$269	—$333	$54.22 plus 31%	—$269
$333		$74.06 plus 35%	—$333

(b) MARRIED person—

If the amount of wages is:		The amount of income tax to be withheld shall be:	
Not over $11		0	
Over—	But not over—		of excess over—
$11	—$39	14%	—$11
$39	—$167	$3.92 plus 16%	—$39
$167	—$207	$24.40 plus 20%	—$167
$207	—$324	$32.40 plus 24%	—$207
$324	—$409	$60.48 plus 28%	—$324
$409	—$486	$84.28 plus 32%	—$409
$486		$108.92 plus 36%	—$486

TABLE 2. BIWEEKLY Payroll Period

(a) SINGLE person—including head of household:

If the amount of wages is:		The amount of income tax to be withheld shall be:	
Not over $21		0	
Over—	But not over—		of excess over—
$21	—$69	14%	—$21
$69	—$146	$6.72 plus 18%	—$69
$146	—$404	$20.58 plus 21%	—$146
$404	—$462	$74.76 plus 23%	—$404
$462	—$538	$88.10 plus 27%	—$462
$538	—$665	$108.62 plus 31%	—$538
$665		$147.99 plus 35%	—$665

(b) MARRIED person—

If the amount of wages is:		The amount of income tax to be withheld shall be:	
Not over $21		0	
Over—	But not over—		of excess over—
$21	—$79	14%	—$21
$79	—$335	$8.12 plus 16%	—$79
$335	—$413	$49.08 plus 20%	—$335
$413	—$648	$64.68 plus 24%	—$413
$648	—$817	$121.08 plus 28%	—$648
$817	—$971	$168.40 plus 32%	—$817
$971		$217.68 plus 36%	—$971

TABLE 3. SEMIMONTHLY Payroll Period

(a) SINGLE person—including head of household:

If the amount of wages is:		The amount of income tax to be withheld shall be:	
Not over $23		0	
Over—	But not over—		of excess over—
$23	—$75	14%	—$23
$75	—$158	$7.28 plus 18%	—$75
$158	—$438	$22.22 plus 21%	—$158
$438	—$500	$81.02 plus 23%	—$438
$500	—$583	$95.28 plus 27%	—$500
$583	—$721	$117.69 plus 31%	—$583
$721		$160.47 plus 35%	—$721

(b) MARRIED person—

If the amount of wages is:		The amount of income tax to be withheld shall be:	
Not over $23		0	
Over—	But not over—		of excess over—
$23	—$85	14%	—$23
$85	—$363	$8.68 plus 16%	—$85
$363	—$448	$53.16 plus 20%	—$363
$448	—$702	$70.16 plus 24%	—$448
$702	—$885	$131.12 plus 28%	—$702
$885	—$1,052	$182.36 plus 32%	—$885
$1,052		$235.80 plus 36%	—$1,052

TABLE 4. MONTHLY Payroll Period

(a) SINGLE person—including head of household:

If the amount of wages is:		The amount of income tax to be withheld shall be:	
Not over $46		0	
Over—	But not over—		of excess over—
$46	—$150	14%	—$46
$150	—$317	$14.56 plus 18%	—$150
$317	—$875	$44.62 plus 21%	—$317
$875	—$1,000	$161.80 plus 23%	—$875
$1,000	—$1,167	$190.55 plus 27%	—$1,000
$1,167	—$1,442	$235.64 plus 31%	—$1,167
$1,442		$320.89 plus 35%	—$1,442

(b) MARRIED person—

If the amount of wages is:		The amount of income tax to be withheld shall be:	
Not over $46		0	
Over—	But not over—		of excess over—
$46	—$171	14%	—$46
$171	—$725	$17.50 plus 16%	—$171
$725	—$896	$106.14 plus 20%	—$725
$896	—$1,404	$140.34 plus 24%	—$896
$1,404	—$1,771	$262.26 plus 28%	—$1,404
$1,771	—$2,104	$365.02 plus 32%	—$1,771
$2,104		$471.58 plus 36%	—$2,104

TABLE 5. QUARTERLY Payroll Period

(a) SINGLE person—including head of household:

If the amount of wages is:		The amount of income tax to be withheld shall be:	
Not over $138		0	
Over—	But not over—		of excess over—
$138	—$450	14%	—$138
$450	—$950	$43.68 plus 18%	—$450
$950	—$2,625	$133.68 plus 21%	—$950
$2,625	—$3,000	$485.43 plus 23%	—$2,625
$3,000	—$3,500	$571.68 plus 27%	—$3,000
$3,500	—$4,325	$706.68 plus 31%	—$3,500
$4,325		$962.43 plus 35%	—$4,325

(b) MARRIED person—

If the amount of wages is:		The amount of income tax to be withheld shall be:	
Not over $138		0	
Over—	But not over—		of excess over—
$138	—$513	14%	—$138
$513	—$2,175	$52.50 plus 16%	—$513
$2,175	—$2,688	$318.42 plus 20%	—$2,175
$2,688	—$4,213	$421.02 plus 24%	—$2,688
$4,213	—$5,313	$787.02 plus 28%	—$4,213
$5,313	—$6,313	$1,095.02 plus 32%	—$5,313
$6,313		$1,415.02 plus 36%	—$6,313

TABLE 6. SEMIANNUAL Payroll Period

(a) SINGLE person—including head of household:

If the amount of wages is:		The amount of income tax to be withheld shall be:	
Not over $275		0	
Over—	But not over—		of excess over—
$275	—$900	14%	—$275
$900	—$1,900	$87.50 plus 18%	—$900
$1,900	—$5,250	$267.50 plus 21%	—$1,900
$5,250	—$6,000	$971.00 plus 23%	—$5,250
$6,000	—$7,000	$1,143.50 plus 27%	—$6,000
$7,000	—$8,650	$1,413.50 plus 31%	—$7,000
$8,650		$1,925.00 plus 35%	—$8,650

(b) MARRIED person—

If the amount of wages is:		The amount of income tax to be withheld shall be:	
Not over $275		0	
Over—	But not over—		of excess over—
$275	—$1,025	14%	—$275
$1,025	—$4,350	$105.00 plus 16%	—$1,025
$4,350	—$5,375	$637.00 plus 20%	—$4,350
$5,375	—$8,425	$842.00 plus 24%	—$5,375
$8,425	—$10,625	$1,574.00 plus 28%	—$8,425
$10,625	—$12,625	$2,190.00 plus 32%	—$10,625
$12,625		$2,830.00 plus 36%	—$12,625

TABLE 7. ANNUAL Payroll Period

(a) SINGLE person—including head of household:

If the amount of wages is:		The amount of income tax to be withheld shall be:	
Not over $550		0	
Over—	But not over—		of excess over—
$550	—$1,800	14%	—$550
$1,800	—$3,800	$175.00 plus 18%	—$1,800
$3,800	—$10,500	$535.00 plus 21%	—$3,800
$10,500	—$12,000	$1,942.00 plus 23%	—$10,500
$12,000	—$14,000	$2,287.00 plus 27%	—$12,000
$14,000	—$17,300	$2,827.00 plus 31%	—$14,000
$17,300		$3,850.00 plus 35%	—$17,300

(b) MARRIED person—

If the amount of wages is:		The amount of income tax to be withheld shall be:	
Not over $550		0	
Over—	But not over—		of excess over—
$550	—$2,050	14%	—$550
$2,050	—$8,700	$210.00 plus 16%	—$2,050
$8,700	—$10,750	$1,274.00 plus 20%	—$8,700
$10,750	—$16,850	$1,684.00 plus 24%	—$10,750
$16,850	—$21,250	$3,148.00 plus 28%	—$16,850
$21,250	—$25,250	$4,380.00 plus 32%	—$21,250
$25,250		$5,660.00 plus 36%	—$25,250

TABLE 8. DAILY or MISCELLANEOUS Payroll Period

(a) SINGLE person—including head of household:

If the wages divided by the number of days in such period are:		The amount of income tax to be withheld shall be the following amount multiplied by the number of days in such period:	
Not over $1.50		0	
Over—	But not over—		of excess over—
$1.50	—$4.90	14%	—$1.50
$4.90	—$10.40	$0.48 plus 18%	—$4.90
$10.40	—$28.80	$1.47 plus 21%	—$10.40
$28.80	—$32.90	$5.33 plus 23%	—$28.80
$32.90	—$38.40	$6.27 plus 27%	—$32.90
$38.40	—$47.40	$7.76 plus 31%	—$38.40
$47.40		$10.55 plus 35%	—$47.40

(b) MARRIED person—

If the wages divided by the number of days in such period are:		The amount of income tax to be withheld shall be the following amount multiplied by the number of days in such period:	
Not over $1.50		0	
Over—	But not over—		of excess over—
$1.50	—$5.60	14%	—$1.50
$5.60	—$23.80	$0.57 plus 16%	—$5.60
$23.80	—$29.50	$3.48 plus 20%	—$23.80
$29.50	—$46.20	$4.62 plus 24%	—$29.50
$46.20	—$58.20	$8.63 plus 28%	—$46.20
$58.20	—$69.20	$11.99 plus 32%	—$58.20
$69.20		$15.51 plus 36%	—$69.20

MECHANICS LIEN LAW

Mechanics Lien laws are promulgated by the individual states. The *lien law* is of critical importance to all persons involved with construction properties; building and real estate owners, their agents, contractors, subcontractors, architects, consultants, surety companies, banks, suppliers, and not in the least; laborers.

Most states have some kind of lien law to provide special protection to persons who furnish labor or materials to improve private real property. If an owner of such property fails to make due payment, that property is subject to a "lien". This means that the property is held in debt until such debt is satisfied. Almost every contractor's license examination in The United States will contain some questions concerning Mechanics' Lien Law. Although the intent of the lien law in all states is basically the same, the language and terminology may vary somewhat. Time schedule for posting, filing or serving, as well as the manner in which these procedures are performed may vary widely. Failure to comply with the legal procedures will result in loss of lien rights.

A comparison of parts of The Mechanics Lien Law between California and Florida is shown in Figures 7.1 and 7.2. In California, The Mechanics Lien Law is part of the California Civil Code, Sections 2395 to 3273. It is not published by the State of California, but may be procured from West Publishing Co., 417 South Hill Street, Los Angeles, for $4.50. In Florida, the lien law is part of the Florida Statutes, Chapter 713. It is published by the State of Florida as a separate pamphlet and is usually available free of charge. Obviously each state is different; for information the candidate should write to his State Capital, Division of Documents.

SUNDRY LAWS

There are several other laws that may form the basis of contractor license examination questions. These are likely to include the following:

> State Licensing Law
> Unemployment Compensation Law
> Workmen's Compensation Law
> Sales and Use Tax Law

All large cities will have a State Office; a telephone call to that office will usually prove fruitful in obtaining a copy of the pertinent laws. If there is no State Office in your city write to the State Capital, Division of Documents.

TIME SCHEDULE FOR POSTING, FILING OR SERVING DOCUMENTS
OF THE MECHANICS' LIEN LAW - FLORIDA

TYPE OF DOCUMENT	PERSON WHO POSTS, SERVES OR FILES	TIME TO POST, SERVE OR FILE
Notice to Owner (713.06)	Anyone entitled to a mechanics lien except laborers	Post a notice on the owner before commencing or not later than 45 days from commencing
Claim of Lien (713.08)	Every lienor including laborers and persons in privity	File in the County Clerk's office not later than 90 days after completion of work
Notice of Delivery (713.09)	Seller and purchaser of materials delivered to a place other than the job site	Serve on owner a notice signed by both seller and purchaser
Notice of Commencement (713.135)	Owner or his authorized Agent	Record in County Clerk's office before actually commencing work.
Waiver or Release (713.20)	Any lienor other than a laborer	Any time
Notice of Contest of Lien (713.22)	Owner or his Agent or Attorney	Record in County Clerk's office
Payment Bond (713.23)	Contractor	Before commencing the construction

FIGURE 7.1

TIME SCHEDULE FOR POSTING, FILING OR SERVING DOCUMENTS
OF THE MECHANICS' LIEN LAW - CALIFORNIA

TYPE OF DOCUMENT	PERSON WHO POSTS, SERVES OR FILES	TIME AND PLACE TO SERVE OR FILE
Ninety-day Public Works Preliminary Bond Notice (3092)	Any claimant other than one who performed actual labor for wages or who has direct contractual relationship with the original contractor.	No later than 90 days from the date on which the claimant furnished the last labor, services, equipment or materials. By registered or certified mail to the original contractor.
Notice of Cessation (30920)	Owner or his authorized agent.	Recorded in the office of the County Recorder after cessation of labor. Effective only if there has been a continuous cessation for at least 30 days prior to such recording.
Notice of Completion (3092	Owner or his authorized agent.	Recorded in the office of the County Recorder within 10 days of such completion.
Notice of Nonresponsibility (3094)	A person owning or claiming an interest in the site who has not caused the improvement to be performed	Within 10 days after obtaining knowledge of the work or improvement, by posting notice in a conspicuous place on the site and recording in the office of the County Recorder.
Preliminary 20-day notice--private work (3097)	Anyone--except a person under direct contract with the owner or performing actual labor for wages-- who furnishes labor, service, equipment of material.	Not later than 20 days after the claimant has first furnished labor, service or equipment or materials to the job site by delivering or mailing to the person to be notified.

FIGURE 7.2

TIME SCHEDULE FOR POSTING, FILING OR SERVING DOCUMENTS
OF THE MECHANICS' LIEN LAW

TYPE OF DOCUMENT	PERSON WHO POSTS, SERVES OR FILES	TIME AND PLACE TO SERVE OR FILE
Claim of Lien (3115)	Original contractor.	File in the office of the County Recorder not later than 90 days after completion of contract.
Claim of Lien (3116)	Other than original contractor.	File in the office of the County Recorder after he has ceased furnishing labor, services, equipment or materials and before 90 days after completion of the improvement.
Surety Bond (3144)	Owner	Record in the office of County Recorder either before or after commencement of action to enforce a claim of lien.
Stop Notice (3103) (3184)	Any claimant or his agent.	Delivered by hand, or registered or certified mail to the owner, his architect, or a construction lender within 30 days after the recording of a notice of completion has been filed, then within 90 days after the actual completion.
Commencement of Actions (3172)	Any claimant or his agent.	After 10 days from the date of service of the Stop Notice and no later than 90 days following the expiration of the period within which claims of liens must be recorded. In the same manner as above.

FIGURE 7.2 (continued)

Because no two state laws are similar, it would be impossible to offer even a cursory coverage of all the state laws in any single volume. Usually, but not always, the laws required for study will be listed in the recommended reference list published by the examining board for each locality. Sometimes, an abstract or synopsis of the law is published in small pamphlet form and is adequate for examination preparation.

8. BASIC MATHEMATICS

FRACTIONS

To Reduce Common Fractions: divide the numerator and
denominator by common divisors until further reduction
is impossible -

$$63/81 = 3\overline{)63}^{\,21} \quad 3\overline{)81}^{\,27} = 21/27 = 7/9$$

To Reduce Improper Fractions: divide the numerator by
the denominator, the quotient being a whole number and
and the remainder the new numerator -

$$43/6 = 6\overline{)43} = 7\text{-}1/6$$

To Express a Fraction as a Decimal: divide the numerator by the denominator -

$$3/4 = 4\overline{)3} = 4\overline{)3.00}^{\,.75} = 0.75$$

To Reduce Complex Fractions: first express both numerator and denominator as simple fractions, then multiply
the upper numerator by the lower denominator for the
new numerator and the lower numerator by the upper
denominator -

$$\frac{1\text{-}3/4}{5/6} = \frac{7/4}{5/6} = 42/20 = 20\overline{)42} = 2\text{-}1/10$$

To Reduce Fractions to a Common Denominator: multiply
the numerator of each fraction by the product of all
of the denominators except its own for the new numerators and multiply all denominators together for the
new common denominator -

$$2/3, 1/4, 3/5 = 40/60, 15/60, 36/60$$

To Add Fractions: reduce to a common denominator and
add the numerators -

$$3/4 + 2/3 = 9/12 + 8/12 = 17/12 = 1\text{-}5/12$$

385

To Subtract Fractions: reduce to a common denominator and subtract the numerators -

3/4 - 2/3 = 9/12 - 8/12 = 1/12

To Multiply Fractions: multiply the numerators for a new numerator and multiply the denominators for a new denominator -

3/4 x 5/8 = 15/32

To Divide Fractions: invert the divisor and multiply -

a. 3/4 ÷ 7/8 = 3/4 x 8/7 = 24/28 = 6/7

b. 1-1/8 ÷ 3/16 = 9/8 ÷ 3/16 = 9/8 x 16/3 = 144/24 =

6/1 = 6

DECIMALS

To Express a Decimal as a Fraction: disregard the decimal point and write the figures as the numerator of the fraction. Write the denominator as 1 plus as many zeros after it as there were figures following the decimal -

$$.0125 = \frac{125}{10000}$$

To Express a Fraction as a Decimal: divide the numerator by the denominator -

$$32/64 = 1/2 = 2\overline{)1.0}^{.5} = .5$$

To Multiply Decimals: proceed as in simple multiplication and point off, as many decimal places in the product as are in the multiplier and multiplicand together -

$87.96 x 23.5 =

```
    87.96   =  3 places
  x 23.5
   43980
   26388
   17592
$ 2067.060 ( 3 places )
```

To Divide Decimals: proceed as in simple division and point off, as many decimal places in the quotient as are in the dividend in excess of the divisor -

$$0.2546 \div 0.38 = \quad .38 \, / \overline{\quad.25\,46\quad}^{\;.67} = \quad 0.67$$

$$\begin{array}{r} 22\;8 \\ \hline 26\;6 \\ 26\;6 \end{array}$$

RATIO AND PROPORTION

Ratio is the relation of one figure to another and is sometimes expressed as a fraction with the first quantity as the numerator:

The ratio of 1 to 2 = 1:2 = 1/2

When ratios are equal to each other they are said to be in proportion. The ratio of 3 to 6 = 3:6 = 1/2, therefore it is equal to and in proportion to the ratio of 1 to 2 and the proportion would be written,

3:6 = 1:2

and, read, " 3 is to 6 as 1 is to 2 ".

The first and last terms in a statement of proportion are called the *extremes*, and the middle terms are the *means*. A rule of proportion is that the product of the extremes is equal to the product of the means; therefore, in the above example;

3:6 = 1:2 = 3 x 2 = 6 and 6 x 1 = 6

When the middle terms are identical this quantity is called the mean proportional of the first and last terms;

1:2 = 2:4, 2 is the mean proportional between 1 and 4. To find the mean proportional of any two terms multiply them and extract the square root of their product. Thus the mean proportional of 2 and 50 is;

$\sqrt{2 \times 50}$ = $\sqrt{100}$ = 10. Therefore, 2:10 = 10:50

If proportion is expressed algebraically as;

a:b = c:d

then given any three terms the fourth can be determined. In direct proportions we may "solve for x" by one of the following formulas:

1.) $a = \dfrac{b\ c}{d}$ 3.) $b = \dfrac{a\ d}{c}$

2.) $c = \dfrac{a\ d}{b}$ 4.) $d = \dfrac{b\ c}{a}$

In the following examples, solve for x by using the above formulas;

1. $24:4$ as $x:3 = c = \dfrac{a\ d}{b}$

   $= \dfrac{24 \times 3}{4} = \dfrac{72}{4} = 18 = 24:4$ as $18:3$

2. $2:3$ as $4:x = d = \dfrac{b\ c}{a}$

   $= \dfrac{3 \times 4}{2} = \dfrac{12}{2} = 6 = 2:3$ as $4:6$

3. $5:x\ 25:20 = b = \dfrac{a\ d}{c}$

   $= \dfrac{5 \times 20}{25} = \dfrac{100}{25} = 4 = 5:4$ as $25:20$

4. $x:25$ as $10:2 = a = \dfrac{b\ c}{d}$

   $= \dfrac{25 \times 10}{2} = \dfrac{250}{2} = 125 = 125:25$ as $10:2$

Direct proportion formulas are useful in solving many problems, for example;

5.   A 20 ft wall, shades 3/4 of a lawn, how much higher must the wall be to shade the entire lawn?

   $20:3/4$ as $x:1 = 20:0.75$ as $x:1 =$

   $\dfrac{20 \times 1}{.75} = 26.6$ 26.6 minus $20 = 6.6$ ft

SQUARE AND SQUARE ROOTS

To Find the Square of any Number: multiply the number by itself -

The superscript^2 denotes *square*, thus 4 squared is written;

$$4^2 = 16$$

, and 256 squared is written;

$$256^2 = 65536$$

To Find the Square Root of any Number: by trial and error, determine which number when multiplied by itself will equal the number given -

The symbol $\sqrt{}$ denotes square root, thus the square root of 16 is written;

$$\sqrt{16} = 4$$

, and the square root of 2302.625 is written;

$$\sqrt{2302.625} = 47.98$$

Finding the square root of numbers of 3 digits or more is a slow and laborious task, the examination candidate should either use a slide rule or refer to a standard table. Table 3.1 gives the square and square root for numbers from 1 through 100.

TABLE 8.1

SQUARES AND SQUARE ROOTS

No.	Square	Square Root	No.	Square	Square Root	No.	Square	Square Root	No.	Square	Square Root
1	1	1.0000	26	676	5.0990	51	2601	7.1414	76	5776	8.7178
2	4	1.4142	27	729	5.1962	52	2704	7.2111	77	5929	8.7750
3	9	1.7321	28	784	5.2915	53	2809	7.8201	78	6084	8.8318
4	16	2.0000	29	841	5.3852	54	2916	7.3485	79	6241	8.8882
5	25	2.2361	30	900	5.4772	55	3025	7.4162	80	6400	8.9443
6	36	2.4495	31	961	5.5678	56	3136	7.4833	81	6561	9.0000
7	49	2.6458	32	1024	5.6569	57	3249	7.5498	82	6724	9.0554
8	64	2.8284	33	1089	5.7446	58	3364	7.6158	83	6889	9.1104
9	81	3.0000	34	1156	5.8310	59	3481	7.6811	84	7056	9.1652
10	100	3.1623	35	1225	5.9161	60	3600	7.7460	85	7225	9.2195
11	121	3.3166	36	1296	6.0000	61	3721	7.8102	86	7396	9.2736
12	144	3.4641	37	1369	6.0828	62	3844	7.8740	87	7569	9.3276
13	169	3.6056	38	1444	6.1644	63	3969	7.9373	88	7744	9.3808
14	196	3.7417	39	1521	6.2450	64	4096	8.0000	89	7921	9.4340
15	225	3.8730	40	1600	6.3246	65	4225	8.0623	90	8100	9.4868
16	256	4.0000	41	1681	6.4031	66	4356	8.1240	91	8281	9.5394
17	289	4.1231	42	1764	6.4807	67	4489	8.1854	92	8464	9.5917
18	324	4.2426	43	1849	6.5574	68	4624	8.2462	93	8649	9.6437
19	361	4.3589	44	1936	6.6332	69	4761	8.3066	94	8836	9.6954
20	400	4.4721	45	2025	6.7082	70	4900	8.3666	95	9025	9.7468
21	441	4.5826	46	2116	6.7823	71	5041	8.4261	96	9216	9.7980
22	484	4.6904	47	2209	6.8557	72	5184	8.4853	97	9409	9.8489
23	529	4.7958	48	2304	6.9282	73	5329	8.5440	98	9604	9.8995
24	576	4.8990	49	2401	7.0000	74	5476	8.6023	99	9801	9.9499
25	625	5.0000	50	2500	7.0711	75	5625	8.6603	100	10000	10.0000

SQUARE ROOT OF FRACTIONS	
Fraction	Square Root
1/8	.3535
1/4	.5000
3/8	.6124
1/2	.7071
5/8	.7906
3/4	.8660
7/8	.9354

FIGURE 8.1

AREAS AND VOLUMES

1. Perimeter of a Rectangle

P = 2L + 2W

2. Area of a Square

A = L x W

3. Area of a Triangle

$$A = \frac{Base \ x \ Height}{2}$$

4. Area of a Circle

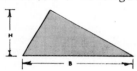

$A = \pi r^2$ (3.142 x Radius squared)

5. Circumference of a Circle

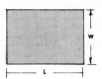

$C = \pi d$ (3.142 x diameter)

FIGURE 8.1 (Continued)

6. Area of a Segment of a Circle

A = area of a circle x number of degrees in arc
    ───
                       360

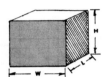

7. Cubical contents of a Rectangle

CC = H x L x W

8. Volume of a Cylinder

V = 0.7854 d² h
(0.7854 x diameter squared x height)

9. USEFUL DATA: ELECTRICAL, HEATING, INSULATION

USEFUL ELECTRICAL DATA

FORMULAE

The following symbols are common usage in electrical formulas:

I	=	Ampere, a unit of current
E	=	Volt, a unit of pressure
W	=	Watt, a measure of power
R	=	Ohm, a unit of resistance
eff	=	Efficiency; use 0.85 when eff is unknown
pf	=	Power factor; the ratio of the actual power to the apparent power
kw	=	1000 watts, a measure of power
hp	=	horsepower; 1 hp = 0.746 kw

Figure 9.1 shows the Power Equation Wheel from which the basic electrical formulas may be read. W may be substituted for P (power). Symbols in the outer ring equal the symbol in the inner ring.

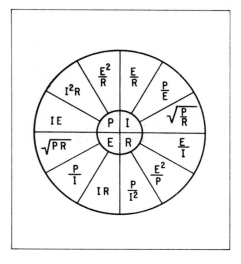

FIGURE 9.1

POWER EQUATION WHEEL

To Find the Power Factor:

1. Single phase circuits; $pf = \dfrac{W}{E \times I}$

2. Two phase circuits; $pf = \dfrac{W}{E \times I \times 2}$

3. Three phase circuits; $pf = \dfrac{W}{E \times I \times 1.73}$

To Find Amps Where Hp is Known:

1. Single phase circuits; $I = \dfrac{hp \times 746}{E \times eff \times pf}$

2. Three phase circuits; $I = \dfrac{hp \times 746}{1.73 \times E \times eff \times pf}$

To Find Hp where Amps are Known:

1. Single phase circuit; $hp = \dfrac{I \times E \times eff \times pf}{746}$

2. Three phase circuit; $hp = \dfrac{I \times E \times eff \times pf \times 1.73}{746}$

THREE PHASE AC CIRCUITS

Three phase is the most common polyphase system. It is connected in either "delta" or "star" (sometimes called Y) formation. Figure 9.2 shows a *delta* system, the current between any two wires is 240 V; the line voltage in a *delta* connection equals the *coil voltage*. Figure 9.3 shows a *star* system, the current between the neutral and any hot wire is single-phase 120 volts; the line voltage in a *star* connection equals the *coil current*.

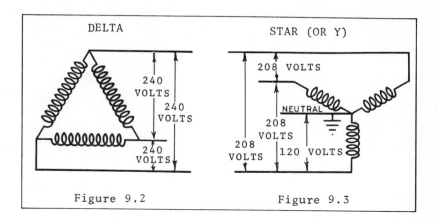

DELTA

STAR (OR Y)

Figure 9.2

Figure 9.3

ELECTRICAL MEASUREMENTS

Electrical units of measurement are given in Table 9.1 together with the instruments generally used to measure each unit. Frequency, inductance (henry), and capacitance (farad) may also be measured but will not be discussed here.

TABLE 9.1

PRACTICAL ABSOLUTE UNITS OF ELECTRICAL QUANTITIES

CODE	DESCRIPTION	SYMBOL	UNIT	MEASURING INSTRUMENT
Voltage (emf)	Pressure	V	Volt	Voltmeter
Resistance		Ω	Ohm	Ohmeter
Current	Flow	I	Ampere	Ammeter
Power	Rate of Energy	W	Watt	Wattmeter
Power Factor	Power Ratio	pf		Power Factor Meter

In Figure 9.4 the application for measuring instruments is shown graphically in the circuit.

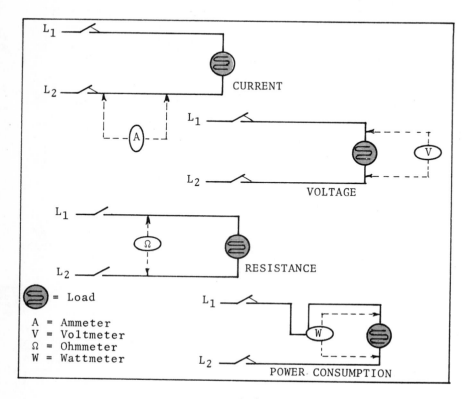

FIGURE 9.4

As a matter of convenience, a prefix system is used to express very large or very small quantities of electrical units, thereby avoiding the use of many zeroes and decimals. Some common prefixes and their equivalents are:

Kilo	=	Thousands	1 kilowatt	=	1000 watts
Meg	=	Millions	1 megohm	=	1,000,000 ohms
Milli	=	1/1000	1 milliamp	=	0.001 amp
Micro	=	1/1,000,000	1 microamp	=	0.000001 amp

MOTOR STARTERS

Figure 9.5 shows a typical line starter diagram. Normally, the "Start" button contacts are open and no current passes through the starter. When the "Start" button is depressed current passes through the contactor, or magnetic coil. When the coil is energized, the plunger "P" is drawn up, magnetically raising the contact bar and completing the circuit between line and fan motor. When the coil is *de-energized* the plunger — released from its magnetic field — drops by gravity and takes the contactor bar with it, thereby breaking the circuit. Releasing the "Start" button does not break the circuit. Once the holding coil is energized the only way to break the circuit is to depress the "Stop" button.

When the temperature rise of a motor is excessive the insulation life is shortened; this is the most serious effect of motor overload. To protect motors against burnout by excessive heat the motor starter usually is equipped with *overload protection*. The most common device is the thermal bi-metallic overload relay installed in the circuit adjacent to the heater coils through which the motor current passes. When abnormally high current passes through the heater coils the overload relays will warp and interrupt the control circuit to the holding coil. See Figure 9.5.

The technician must check the rating of the installed heater elements against the recommended rating number — usually found on the inside cover of the starter. It is not unusual to find improperly sized heater elements in motor starters.

Upon startup, it may be found that the fan motor is rotating in the wrong direction. To reverse a three-phase machine it is only necessary to interchange two of the incoming leads. In the case of single-phase motors, it is necessary to reverse the starting winding relative to the running winding, therefore any links between the starting and running terminals should be removed.

Motors must, of course, be matched to the power supply. Ordinarily the motor *nameplate* gives the name of the manufacturer, rating in horsepower, power supply frequency and voltage, full-load amperes, locked-rotor, temperature rise, service factor, duty cycle and manufacturer's identification number.

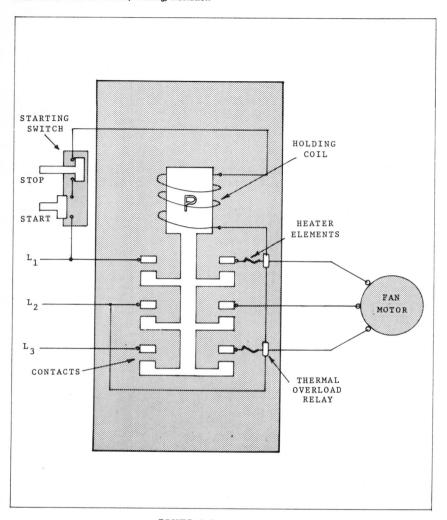

FIGURE 9.5

MOTOR STARTER

Types of Enclosures

Surrounding atmospheric conditions determine the selection of motors. The more enclosed a motor, the more it costs and the hotter it runs.

Open-type: Allows maximum ventilation, has full opening in the frame and end bells.

Semi-protected: Openings are lined with screens to protect the motor from falling particles.

Drip-proof: The upper parts of the motor are shielded to prevent vertical overhead dripping.

Splash-proof: The bottom parts of the motor are shielded to prevent splashing or particles at angle not over 100 deg from vertical.

Totally-enclosed: Non-ventilated, may be used in hazardous atmospheres, or may be explosion proof.

GLOSSARY OF TERMS

Ampere (I): Unit of electrical current produced in a conductor by the applied voltage. One ampere is equal to the flow of one coulomb passing a point in a circuit every second.

Brake Horsepower (bhp): The horsepower *actually* required to drive a fan; it includes the energy losses in the fan but does not include the drive loss between the motor and the fan. The name derives from the Prony brake, a common method of testing mechanical output of motors.

Coulomb (Q): The basic quantity of electrical measurement. One coulomb equals 6.25 billion billion electrons or one ampere per second.

Electromotive Force (EMF): The force that causes current to flow, i.e., the electric potential difference between the terminals of any device used as a source of electrical energy.

Frequency: The number of complete cycles of alternating current per second.

Horsepower (hp): Work performed equivalent to 33,000 foot pounds per minute. The term horsepower was first used by James Watt to measure the power of his steam engine against the power of a London draught horse.

Motor Efficiency (EFF): The rate at which a motor converts electrical energy. The greater the efficiency the lower the energy cost. A well designed induction motor may have an EFF between 75% and 95% of full load.

Ohm (R): A unit of resistance to the flow of current.

Power Factor (PF): The ratio of actual to apparent power of an alternating current found by measuring the current with a watt-meter as against the indicated volt meter reading. It is a measure of the loss in an insulator, capacitor, or inductor.

Service Factor: A safety factor designed into a motor to deliver more than its rated horsepower. A 10 hp motor with a service factor of 1.10 is capable of delivering 11 horsepower.

Temperature Rise: Part of the electrical specification of a motor; a measure of the motor's capacity to dissipate the heat generated by its electrical power losses. If the nameplate shows a temperature rise of 40 C and the ambient temperature is 90 F (32.2C), the motor could operate at 162 F (72C). If the motor becomes hot, hang one thermometer (0-110 C or 0-230 F) in the room near the motor, and place one thermometer in firm contact with the bearing bushes, stator, or other *hot* stationary part of the motor, and check the rise.

Torque: A rotating force. *Starting torque* - also known as locked-rotor or breakaway torque - is the torque available to start a load from a standstill, i.e., the torque exerted by the starting current of the motor to overcome the static friction at rest.

Volt (V or E): A unit of electrical pressure that will push one ampere through a resistance of one ohm.

Watt (W): A unit of power, being the amount of energy expended for one ampere to flow through one ohm.

USEFUL HEATING INFORMATION

PROPERTIES OF STEAM

Water is commonly thought to boil at 212 F (100 C), however, this is true only at standard atmospheric pressure of 14.7 psia (30 in. Hg). By placing the liquid in a closed vessel, and raising or lowering the pressure the water can be vaporized at another temperature. For example, water can be brought to boil at 101 F (38 C) by lowering the pressure to 1 psia; or it can be heated to 353 F (178 C) before it will boil by increasing the pressure to 140 psia.

These data may be found in such books as "Thermodynamic Properties of Steam" by J.H. Keenan and F.G. Keyes, Wiley, N.Y. 1937. For the purpose of contractor examinations, the abridged "Steam Tables" found on pp 358-359 of the *Trane Air Conditioning Manual*, should be quite adequate. Pressure read on a gage is called gage pressure, pressure read on a barometer is absolute pressure. "Steam Table" values are always given in psia (pounds per sq in. absolute). To find the psia in a vessel the pressure of the surrounding atmosphere must be added to the psig (pounds per sq in. gage). Thus, if the gage pressure of the vessel reads 43 psig, and a barometer shows 14.7 psia, then the absolute pressure in the vessel is 43 + 14.7 = 57.7 psia. Each pressure in the table has a corresponding boiling point. The table also gives tabulated values for the specific volume and enthalpy of the saturated liquid and saturated vapor.

GLOSSARY OF TERMS

Enthalpy: The enthalpy of water or steam is the heat that must be added - Btu/lb - to bring it from a liquid at 32 F to its present temperature pressure and condition. The enthalpy of evaporation is the enthalpy difference between saturated liquid and dry saturated vapor.

Superheated Steam: Steam at a temperature higher than the saturation temperature for the given pressure. Superheating is achieved by passing saturated steam through tube coils exposed to furnace heat.

Btu: British thermal unit, the amount of heat required to raise the temperature of one pound of water through one degree Fahrenheit.

Specific Heat: The amount of heat required to raise the temperature of one pound of a substance through one degree Fahrenheit. Typical specific heats are; concrete 0.270, iron 0.13, water 1.0.

EDR: Equivalent Direct Radiation, the amount of surface (equivalent sq ft) that emits 240 Btu/hr at a steam temperature of 215 F and a room temperature of 70 F. Substituting for hot water, the emission rate is 150 Btu/hr.

IBR: Institute of Boiler and Radiator Manufacturers. Heating boilers and equipment are often rated by IBR. IBR Code.

SBI: Steel Boiler Institute. Heating boilers and equipment are often rated by SBI. SBI code.

ASME: American Society of Mechanical Engineers. Code writers for heating boilers. ASME Code.

Boiler data: The nameplate data to be shown on every boiler. The minimum required data are:
1. Manufacturer's name and address.
2. Boiler number and type.
3. SBI symbol.
4. Heating surface in sq ft.
5. ASME Code symbol.

Boiler fittings: Devices such as, valves and gages, permanently attached to the boiler to insure safe and efficient operation. These include;
1. Safety valve to relieve excess boiler steam and prevent explosion.
2. Steam gage to indicate steam pressure (psig) in the boiler.
3. Water gage to show the water level in boiler.
4. Try cocks to test water level.
5. Stop valve to control steam flow from the boiler.
6. Blowoff connection to drain boiler and to reduce the concentration of impurities in the water.
7. Fusible plug to protect against "low water."
8. Feed connection to control feed water to boiler. This includes a stop valve, check valve, and interconnecting piping.

Siphon: A device to prevent steam from entering the steam gage. The ASME Code requires that each boiler must have a siphon with a 1/4 in. pipe connection through which a test gage may be connected for checking the boiler steam gage while the boiler is in operation.

Safety valve: (see boiler fittings) A direct spring-loaded pop type device that opens to prevent boiler pressure from rising above a predetermined level.

Each boiler must have at least one safety valve. If
the boiler has more than 500 sq ft of heating surface
it should have two or more safety valves.

Blowdown: The difference between the opening and closing
pressure settings on the safety valve.

Blowoff: (see boiler fittings) A quick-opening or slow-
opening valve or plug cock.. Ordinary glove or gate
valves should not be used. The ASME Code requires
that all boilers operating at over 100 psi shall have
two blowoff valves on each blowoff pipe.

Steam trap: A steam trap is a device that removes air,
gas and condensate from any steam line, and holds
back the steam. It may be considered as a fast-
acting, automatic condensate valve. Some types of
steam traps:

1. Float trap; ball float rises as condensate fills
the trap body and opens the discharge valve.
2. Open bucket trap; as condensate fills the trap
body, the open bucket floats thus rising and
closing the valve. Condensate continues to enter,
spills over the top of the bucket causing it to
sink. This opens the discharge valve. When the
bucket is emptied it floats again and closes the
valve.
3. Inverted bucket trap; similar to the open bucket,
but the opening is in the bottom of the bucket.
When condensate flows into the trap the bucket
remains down and the valve open. As steam begins
to enter the bucket floats and rises to close the
valve.
4. Thermostatic trap; bellows operated device res-
ponding to temperature changes. Bellows contracts
to open the valve.
5. Float and thermostatic trap; a combination of the
float ball and thermostatic bellows allowing great-
eter air handling ability, especially at start-up.

Boiler horsepower: A measure of the capacity of a boiler
(BHP) not related to mechanical energy, as brake horsepower (Bhp). BHP is defined as the evaporation of
34.5 lb of water per hour from a temperature of 212 F
into dry saturated steam at the same temperature. This
is equivalent to 33,475 Btu/hr or 139.5 sq ft of EDR
steam, or 223.1 sq ft EDR water.

Combustion air: The air required for boiler combustion. For every 4200 Btu boiler gross output, one cfm of combustion air should be provided. For every 17,000 Btu boiler gross output, one cfm of *ventilation* air should be provided.

TABLE 9.2

CONVERSION RATIOS AND EQUIVALENTS FOR HEATING

Multiply	X	To Obtain
Sq ft EDR steam	240	Btu per hr
Sq ft EDR water	160	Btu per hr
Boiler hp	33.5	MBtu per hr
Boiler hp	140	Sq ft EDR steam
Boiler hp	223	Sq ft EDR water
Boiler hp	34.5	Lb per hr steam
Lb per hr steam	970	Btu per hr
Sq ft EDR steam	0.247	Lb per hr steam
Sq ft EDR water	0.155	Lb per hr steam
TO OBTAIN ABOVE	DIVIDE BY ABOVE	STARTING WITH ABOVE

USEFUL INSULATION INFORMATION

Whether insulation is used for cold storage walls, chilled water lines, steam lines, residential attics, or curtain walls, it is based on the principle of slowing down the rate of flow of heat through a substance. This heat transmission depends on the material and thickness of insulation thus constituting a condition of thermal conductivity or transfer coefficient.

GLOSSARY OF TERMS

> k factor: The amount of heat in Btu's transmitted through 1 sq ft of 1 in. thick material for a difference of 1 F per hour from surface to surface. See Figure 9.6.

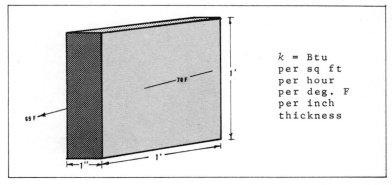

k = Btu
per sq ft
per hour
per deg. F
per inch
thickness

FIGURE 9.6

THERMAL CONDUCTIVITY, k

The k factor for mineral wool blankets is 0.27; therefore, the transference of heat through a 1 in. thick mineral wool blanket would be 0.27 Btu/hr per sq ft per degree difference between the two surfaces.

C factor: Thermal conductance, the amount of heat
in Btu's that will pass through 1 sq ft
per 1 hr between the two surfaces of any
material, or combination of materials
of construction being considered, for a
difference of 1 F; *not per inch of thickness.* The heating, air conditioning,
and refrigerating engineer deals mainly
with compound walls of more than one material, and makes considerable use of conductance factor. Figure 9.7 illustrates
the C factor.

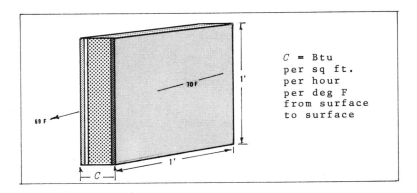

C = Btu
per sq ft.
per hour
per deg F
from surface
to surface

FIGURE 9.7

THERMAL CONDUCTIVITY, C

U factor: Sometimes called the U value, designates the total or overall transmission of heat in Btu's in 1 hr per sq ft of area for a difference in temperature of 1 deg F between the air on one side to the air on the other side of any construction. The term is applied to the usual combinations of materials for walls, roofs, floors or ceiling, as well as for single materials such as glass windows. Figure 9.8 illustrates the U factor.

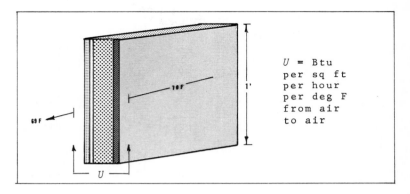

U = Btu per sq ft per hour per deg F from air to air

FIGURE 9.8

OVERALL COEFFICIENT, "U"

R: Thermal resistance, the reciprocal of k, C, or U expressed as degree F per Btu per 1 hr per sq ft. The reciprocal of any number is 1 divided by the number. Thus, a wall with a U value of 0.37 will have a resistance of 2.7

$$\frac{1}{0.37} = 2.7 \; R$$

Industrial insulation: Includes the application of thermal insulation to cold storage rooms, pipes, tanks and vessels, and ducts as opposed to general building insulation.

Water vapor barrier: A material that will reduce the rate of volume of water vapor under specific conditions. Water vapor in air is a gas, which can move through a material under differences in its own vapor pressure independently of the air with which it is mixed.

Surface condensation: The condensation that may appear on the warm side of insulated pipes, ducts, rooms, etc., when the water vapor in the air comes in contact with a surface whose temperature is lower than the dewpoint of the air. A sufficient thickness of insulation should be used to keep the temperature of the surface of the insulation higher than the dewpoint temperature.

Calculating Overall Coefficients

Tables of Overall Coefficient *U* were discussed in Chapter 4, section, *Load Estimating*. Usually, the designer will work from these tables, but frequently he must calculate the Overall Coefficient for a combination of materials not listed in the tables.

To calculate the *U* value of a floor, wall, ceiling or roof it is necessary to know the *k* or *C* of all of the component material making up the section. The conductivities and/or conductances are not additive, but must be converted to resistances. Values for various materials may be found in the ASHRAE *Handbook of Fundamentals, 1972*, Chapter 20. Once the values for *k* or *C* have been determined and the resistances are calculated, the *U* value may be found by the formula:

$$U = \frac{1}{R_1 + R_2 + R_3 + R_4 + \ldots + R_n}$$

Example: Find the *U* value of a frame wall consisting of 1/2 in. gypsum board, 3-1/3 in. mineral wool blanket, 3/4 in. plywood, building paper, and wood shingles.

Solution:

Section	Resistance R	
Air film (outside, 15 mph)	1/6.0	= 0.17
Gypsum board, 1/2 in.	1/2.25	= 0.44
Mineral wool, 3-1/2 in.	1/0.09	=11.00
Plywood, 3/4 in.	1/1.07	= 0.93
Building paper	1/8.35	= 0.12
Wood shingles	1/1.06	= 0.94
Air film (inside still air)	1/1.46	= 0.68
Total resistance of each sq ft		=14.28

(see *ASHRAE H.B. Fundamentals*, 1972, p.349)

Therefore, $U = \dfrac{1}{R_1 + R_2 + R_3 + R_4 + R_5 + R_6 + R_7}$

$= \dfrac{1}{14.28} = 0.07$ Btu per hr per sq ft per deg F

Now, if the wall used in the above example was 30 ft
x 10 ft and the temperature difference between the
inside and outside was 20 F, then:

300 sq ft x 20 F x .07 U = 420 Btuh heat transmission

The example is illustrated graphically in Figure 9.9.

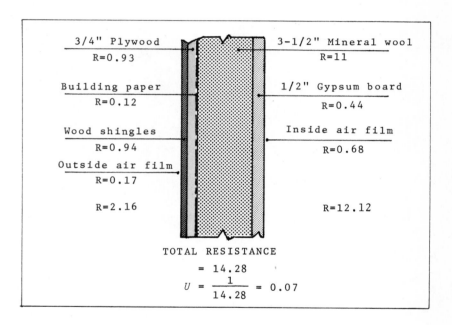

3/4" Plywood
R=0.93

Building paper
R=0.12

Wood shingles
R=0.94

Outside air film
R=0.17

R=2.16

3-1/2" Mineral wool
R=11

1/2" Gypsum board
R=0.44

Inside air film
R=0.68

R=12.12

TOTAL RESISTANCE
= 14.28
$U = \dfrac{1}{14.28} = 0.07$

FIGURE 9.9
CALCULATING COEFFICIENT U

APPENDIX

TABLE A1

METRIC CONVERSION

	Multiply	By	To Obtain	
Length	Inches	25.4	Millimeters	(mm)
	Inches	2.5	Centimeters	(cm)
	Feet	0.305	Meters	(m)
	Feet	30.5	Centimeters	(cm)
Energy	Btu	0.252	Kilogram-Calories	(kcal)
	Btu	107.6	Kilogram-Meters	(kcm)
	Tons (12,000 Btu)	3.024	Kilocalories per hr	(kcal/h)
Area	Square Inches	6.5	Square Centimeters	(cm^2)
	Square Feet	0.09	Square Meters	(m^2)
	Square Yards	0.8	Square Meters	(m^2)
Flow	Cubic feet/minute	1.7	Cubic Meters per hr	(m^3/h)
	Cubic feet/minute	472	Cubic Centimeters per second	(cm/sec)
	Gallons per minute	0.0631	Liters per second	(l/s)
Volume	Cubic Inches	16.4	Cubic Centimeters	(cm^3)
	Cubic Feet	0.028	Cubic Meters	(m^3)
Pressure	Pounds per Square Inch	0.0703	Kilogram per Sq. Centimeter	(kg/cm^2)
	Feet of Water	0.0304	Kilogram per Sq. Centimeter	(kg/cm^2)
Motion	Feet per Minute	0.005	Meters per Second	(m/sec)
	Feet per Minute	0.508	Centimeters per Second	(cm/sec)
	Feet per Minute	1.6903	Kilometers per Hour	(km/h)
Temperature	$^\circ F = (\frac{9}{5}\,^\circ C + 32^\circ)$		$^\circ C = \frac{5}{9}\,(^\circ F - 32^\circ)$	

TABLE A2

USEFUL CONVERSION FACTORS		
Multiply	By	To Obtain
Atmosphere	29.92	Inches of mercury
Atmosphere	33.93	Feet of water
Atmosphere	14.70	Pounds per sq in.
Atmosphere	1.058	Tons per square foot
Barrels (oil)	42	Gallons
Boiler horsepower	33,475	Btu per hour
Boiler horsepower	34.5	Pounds water evaporated from and at 212F
Btu	778	Foot-pounds
Btu	0.000393	Horsepower-hours
Btu	0.000293	Kilowatt-hours
Btu	0.0010307	Pounds water evaporated from and at 212F
Btu per 24 hr	0.00000347	Tons of refrigeration
Btu per hour	0.00002986	Boiler horsepower
Btu per hour	0.000393	Horsepower
Btu per hour	0.000293	Kilowatts
Btu per inch per sq ft per hr per F.	0.0833	Btu per foot per sq ft per hour per F
Cubic feet	1,728	Cubic inches
Cubic feet	7.48052	Gallons
Cubic feet of water	62.37	Pounds (at 60F)
Cubic feet per minute	0.1247	Gallons per second
Feet of water	0.881	Inches of mercury (at 32F)
Feet of water	62.37	Pounds per sq ft
Feet of water	0.4335	Pounds per sq in
Feet of water	0.02950	Atmospheres
Feet per minute	0.01136	Miles per hour
Feet per minute	0.01667	Feet per second
Foot-pounds	0.001286	Btu
Gallons (U.S.)	0.1337	Cubic feet
Gallons (U.S.)	231	Cubic inches
Gallons of water	8.3453	Pounds of water (at 60F)
Horsepower	550	Foot-pounds per sec
Horsepower	33,000	Foot-pounds per min
Horsepower	2,546	Btu per hour
Horsepower	42.42	Btu per minute
Horsepower	0.7457	Kilowatts
Horsepower (boiler)	33,475	Btu per hour

TABLE A2 (Continued)

Multiply	By	To Obtain
Inches of mercury (at 62F)	13.57	In. of Water (at 62F)
Inches of mercury (at 62F)	1.131	Ft of water (at 62F)
Inches of mercury (at 62F)	70.73	Pounds per sq ft
Inches of mercury (at 62F)	0.4912	Pounds per sq in
Inches of water (at 62F)	0.07355	Inches of mercury
Inches of water (at 62F)	0.03613	Pounds per sq in
Inches of water (at 62F)	5.202	Pounds per sq ft
Inches of water (at 62F)	0.002458	Atmospheres
Kilowatts	56.92	Btu per minute
Kilowatts	1.341	Horsepower
Kilowatt-hours	3415	Btu
Latent heat of ice	143.33	Btu per pound
Pounds	7,000	Grains
Pounds of water (at 60F)	0.01602	Cubic feet
Pounds of water (at 60F)	27.68	Cubic inches
Pounds of water (at 60F)	0.1198	Gallons
Pounds of water evaporated from and at 212F	0.284	Kilowatt-hours
Pounds of water evaporated from and at 212F	0.381	Horsepower-hours
Pounds of water evaporated from and at 212F	970.4	Btu
Pounds per square inch	2.0416	In. of mercury
Pounds per square inch	2.31	Ft of water (at 62F)

TABLE A3

INCH--FOOT--DECIMAL CONVERSION

Inches, Fractions	Inches, Decimals	Feet, Decimals	Inches, Fractions	Inches, Decimals	Feet, Decimals	Inches, Fractions	Inches, Decimals	Feet, Decimals
1/64	.0156	.0013	11/32	.3438	.0287	43/64	.6719	.0560
1/32	.0313	.0026	23/64	.3594	.0299	11/16	.6875	.0573
3/64	.0469	.0039				45/64	.7031	.0586
1/16	.0625	.0052	3/8	.3750	.0313	23/32	.7188	.0599
5/64	.0781	.0065	25/64	.3906	.0326	47/64	.7344	.0612
3/32	.0938	.0078	13/32	.4063	.0339			
7/64	.1094	.0091	27/64	.4219	.0352	3/4	.7500	.0625
			7/16*	.4375	.0365	49/64	.7656	.0638
1/8	.1250	.0104	29/64	.4531	.0378	25/32	.7813	.0651
9/64	.1406	.0117	15/32	.4688	.0391	51/64	.7969	.0664
5/32	.1563	.0130	31/64	.4844	.0404	13/16	.8125	.0677
11/64	.1719	.0143				53/64	.8281	.0690
3/16	.1875	.0156	1/2	.5000	.0417	27/32	.8437	.0703
13/64	.2031	.0169	33/64	.5156	.0430	55/64	.8594	.0716
7/32	.2188	.0182	17/32	.5313	.0443			
15/64	.2343	.0195	35/64	.5469	.0456	7/8	.8750	.0729
			9/16	.5625	.0469	57/64	.8906	.0742
1/4	.2500	.0208	37/64	.5781	.0482	29/32	.9063	.0755
17/64	.2656	.0221	19/32	.5938	.0495	59/64	.9219	.0768
9/32	.2813	.0234	39/64	.6094	.0508	15/16	.9375	.0781
19/64	.2969	.0247	5/8	.6250	.0521	61/64	.9531	.0794
5/16	.3125	.0260	41/64	.6406	.0534	31/32	.9688	.0807
21/64	.3281	.0273	21/32	.6563	.0547	63/64	.9844	.0820

TABLE A4

DECIMALS OF A FOOT

In Frac.	0"	1"	2"	3"	4"	5"	6"	7"	8"	9"	10"	11"
0	.000	.083	.167	.250	.333	.417	.500	.583	.667	.750	.833	.917
1/8"	.010	.094	.177	.260	.344	.427	.510	.594	.677	.760	.844	.927
1/4"	.021	.104	.188	.271	.354	.438	.521	.604	.688	.771	.854	.938
3/8"	.031	.115	.198	.281	.365	.448	.531	.615	.698	.781	.865	.948
1/2"	.042	.125	.208	.292	.375	.458	.542	.625	.708	.792	.875	.958
5/8"	.052	.135	.219	.302	.385	.469	.552	.635	.719	.802	.885	.969
3/4"	.063	.146	.229	.313	.396	.479	.563	.646	.729	.813	.896	.979
7/8"	.073	.156	.240	.323	.406	.490	.573	.656	.740	.823	.906	.990

To change decimals of a foot to inches, multiply the decimal by 12.

To change inches to decimals of a foot, divide inches by 12.

TABLE A5

SATURATED REFRIGERANT
Temperature - Pressure Chart

Italicized figures are inches of mercury;
bold type is gauge pressure in lbs per sq in.

Temp. °F	R-717 (Ammonia)	R-11	R-12	R-22	R-500	R-502
	psig	psig	psig	psig	psig	psig
-70	*21.9*	*29.5*	*21.8*	*16.6*	*20.3*	*12.6*
-65	*20.4*	*29.3*	*20.5*	*14.4*	*18.8*	*10.0*
-60	*18.6*	*29.2*	*19.0*	*12.0*	*17.0*	*7.0*
-55	*16.6*	*29.0*	*17.3*	*9.2*	*15.0*	*3.6*
-50	*14.3*	*28.9*	*15.4*	*6.2*	*12.8*	**0.0**
-45	*11.7*	*28.7*	*13.3*	*2.7*	*10.4*	**2.1**
-40	*8.7*	*28.4*	*10.9*	**0.5**	*7.6*	**4.3**
-35	*5.4*	*28.1*	*8.3*	**2.6**	*4.6*	**6.7**
-30	*1.6*	*27.8*	*5.5*	**4.8**	*1.2*	**9.4**
-25	**1.3**	*27.4*	*2.3*	**7.4**	**1.2**	**12.3**
-20	**3.6**	*27.0*	**0.6**	**10.2**	**3.2**	**15.5**
-15	**6.2**	*26.5*	**2.5**	**13.2**	**5.4**	**19.0**
-10	**9.0**	*26.0*	**4.5**	**16.5**	**7.8**	**22.8**
-5	**12.2**	*25.4*	**6.7**	**20.1**	**10.4**	**26.7**
0	**15.7**	*24.7*	**9.2**	**24.0**	**13.3**	**31.2**
5	**19.6**	*23.9*	**11.8**	**28.2**	**16.4**	**36.0**
10	**23.8**	*23.1*	**14.6**	**32.8**	**19.7**	**41.1**
15	**28.4**	*22.1*	**17.7**	**37.7**	**23.4**	**46.6**
20	**33.5**	*21.1*	**21.0**	**43.0**	**27.3**	**52.5**
25	**39.0**	*19.9*	**24.6**	**48.8**	**31.5**	**58.7**
30	**45.0**	*18.6*	**28.5**	**54.9**	**36.0**	**65.4**
35	**51.6**	*17.2*	**32.6**	**61.5**	**40.9**	**72.6**
40	**58.6**	*15.6*	**37.0**	**68.5**	**46.1**	**80.2**
45	**66.3**	*13.9*	**41.6**	**76.0**	**51.7**	**88.3**
50	**74.5**	*12.0*	**46.7**	**84.0**	**57.6**	**96.9**
55	**83.4**	*10.0*	**52.0**	**92.6**	**63.9**	**106.0**
60	**92.9**	*7.8*	**57.7**	**101.6**	**70.6**	**115.6**
65	**103.1**	*5.4*	**63.8**	**111.2**	**77.8**	**125.8**
70	**114.1**	*2.8*	**70.2**	**121.4**	**85.4**	**136.6**
75	**125.8**	**0.0**	**77.0**	**132.3**	**93.5**	**148.0**
80	**138.3**	**1.5**	**84.2**	**143.6**	**102**	**159.9**
85	**151.7**	**3.2**	**91.6**	**155.7**	**111**	**172.6**
90	**165.9**	**4.9**	**99.8**	**168.4**	**121**	**185.8**
95	**181.1**	**6.8**	**108.3**	**181.8**	**131**	**199.8**
100	**197.2**	**8.8**	**117.2**	**195.9**	**141**	**214.4**
105	**214.2**	**10.9**	**126.6**	**210.8**	**153**	**229.8**
110	**232.3**	**13.2**	**136.4**	**226.4**	**164**	**245.8**
115	**251.5**	**15.6**	**146.8**	**242.7**	**176**	**262.7**
120	**271.7**	**18.3**	**157.6**	**259.9**	**189**	**280.3**

TABLE A6

TEMPERATURE CONVERSION FAHRENHEIT-CELSIUS (CENTIGRADE)

In the center column, find the temperature to be converted. The equivalent temperature is in the left column if converting to Celsius, and in the right column if converting to Fahrenheit.

459.4 to 25

C	(value)	F
−273	−459.4	
−268	−450	
−262	−440	
−257	−430	
−251	−420	
−246	−410	
−240	−400	
−234	−390	
−229	−380	
−223	−370	
−218	−360	
−212	−350	
−207	−340	
−201	−330	
−196	−320	
−190	−310	
−184	−300	
−179	−290	
−173	−280	
−169	−273	−459.4
−168	−270	−454
−162	−260	−436
−157	−250	−418
−151	−240	−400
−146	−230	−382
−140	−220	−364
−134	−210	−346
−129	−200	−328
−123	−190	−310
−118	−180	−292
−112	−170	−274
−107	−160	−256
−101	−150	−238
−96	−140	−220
−90	−130	−202
−84	−120	−184
−79	−110	−166
−73	−100	−148
−68	−90	−130
−62	−80	−112
−57	−70	−94
−51	−60	−76
−46	−50	−58
−40	−40	−40
−34	−30	−22
−29	−20	−4
−23	−10	14
−17.8	0	32
−17.2	1	33.8
−16.7	2	35.6
−16.1	3	37.4
−15.6	4	39.2
−15.0	5	41.0
−14.4	6	42.8
−13.9	7	44.6
−13.3	8	46.4
−12.8	9	48.2
−12.2	10	50.0
−11.7	11	51.8
−11.1	12	53.6
−10.6	13	55.4
−10.0	14	57.2
−9.4	15	59.0
−8.9	16	60.8
−8.3	17	62.6
−7.8	18	64.4
−7.2	19	66.2
−6.7	20	68.0
−6.1	21	69.8
−5.6	22	71.6
−5.0	23	73.4
−4.4	24	75.2
−3.9	25	77.0

26 to 99

C	(value)	F
−3.3	26	78.8
−2.8	27	80.6
−2.2	28	82.4
−1.7	29	84.2
−1.1	30	86.0
−0.6	31	87.8
0.0	32	89.6
0.6	33	91.4
1.1	34	93.2
1.7	35	95.0
2.2	36	96.8
2.8	37	98.6
3.3	38	100.4
3.9	39	102.2
4.4	40	104.0
5.0	41	105.8
5.6	42	107.6
6.1	43	109.4
6.7	44	111.2
7.2	45	113.0
7.8	46	114.8
8.3	47	116.6
8.9	48	118.4
9.4	49	120.2
10.0	50	122.0
10.6	51	123.8
11.1	52	125.6
11.7	53	127.4
12.2	54	129.2
12.8	55	131.0
13.3	56	132.8
13.9	57	134.6
14.4	58	136.4
15.0	59	138.2
15.6	60	140.0
16.1	61	141.8
16.7	62	143.6
17.2	63	145.4
17.8	64	147.2
18.3	65	149.0
18.9	66	150.8
19.4	67	152.6
20.0	68	154.4
20.6	69	156.2
21.1	70	158.0
21.7	71	159.8
22.2	72	161.6
22.8	73	163.4
23.3	74	165.2
23.9	75	167.0
24.4	76	168.8
25.0	77	170.6
25.6	78	172.4
26.1	79	174.2
26.7	80	176.0
27.2	81	177.8
27.8	82	179.6
28.3	83	181.4
28.9	84	183.2
29.4	85	185.0
30.0	86	186.8
30.6	87	188.6
31.1	88	190.4
31.7	89	192.2
32.2	90	194.0
32.8	91	195.8
33.3	92	197.6
33.9	93	199.4
34.4	94	201.2
35.0	95	203.0
35.6	96	204.8
36.1	97	206.6
36.7	98	208.4
37.2	99	210.2

100 to 820

C	(value)	F
38	100	212
43	110	230
49	120	248
54	130	266
60	140	284
66	150	302
71	160	320
77	170	338
82	180	356
88	190	374
93	200	392
99	210	410
100	212	413.6
104	220	428
110	230	446
116	240	464
121	250	482
127	260	500
132	270	518
138	280	536
143	290	554
149	300	572
154	310	590
160	320	608
166	330	626
171	340	644
177	350	662
182	360	680
188	370	698
193	380	716
199	390	734
204	400	752
210	410	770
216	420	788
221	430	806
227	440	824
232	450	842
238	460	860
243	470	878
249	480	896
254	490	914
260	500	932
266	510	950
271	520	968
277	530	986
282	540	1004
288	550	1022
293	560	1040
299	570	1058
304	580	1076
310	590	1094
316	600	1112
321	610	1130
327	620	1148
332	630	1166
338	640	1184
343	650	1202
349	660	1220
354	670	1238
360	680	1256
366	690	1274
371	700	1292
377	710	1310
382	720	1328
388	730	1346
393	740	1364
399	750	1382
404	760	1400
410	770	1418
416	780	1436
421	790	1454
427	800	1472
432	810	1490
438	820	1508

830 to 1540

C	(value)	F
443	830	1526
449	840	1544
454	850	1562
460	860	1580
466	870	1598
471	880	1616
477	890	1634
482	900	1652
488	910	1670
493	920	1688
499	930	1706
504	940	1724
510	950	1742
516	960	1760
521	970	1778
527	980	1796
532	990	1814
538	1000	1832
543	1010	1850
549	1020	1868
554	1030	1886
560	1040	1904
566	1050	1922
571	1060	1940
577	1070	1958
582	1080	1976
588	1090	1994
593	1100	2012
599	1110	2030
604	1120	2048
610	1130	2066
616	1140	2084
621	1150	2102
627	1160	2120
632	1170	2138
638	1180	2156
643	1190	2174
649	1200	2192
654	1210	2210
660	1220	2228
666	1230	2246
671	1240	2264
677	1250	2282
682	1260	2300
688	1270	2318
693	1280	2336
699	1290	2354
704	1300	2372
710	1310	2390
716	1320	2408
721	1330	2426
727	1340	2444
732	1350	2462
738	1360	2480
743	1370	2498
749	1380	2516
754	1390	2534
760	1400	2552
766	1410	2570
771	1420	2588
777	1430	2606
782	1440	2624
788	1450	2642
793	1460	2660
799	1470	2678
804	1480	2696
810	1490	2714
816	1500	2732
821	1510	2750
827	1520	2768
832	1530	2786
838	1540	2804

1550 to 2260

C	(value)	F
843	1550	2822
849	1560	2840
854	1570	2858
860	1580	2876
866	1590	2894
871	1600	2912
877	1610	2930
882	1620	2948
888	1630	2966
893	1640	2984
899	1650	3002
904	1660	3020
910	1670	3038
916	1680	3056
921	1690	3074
927	1700	3092
932	1710	3110
938	1720	3128
943	1730	3146
949	1740	3164
954	1750	3182
960	1760	3200
966	1770	3218
971	1780	3236
977	1790	3254
982	1800	3272
988	1810	3290
993	1820	3308
999	1830	3326
1004	1840	3344
1010	1850	3362
1016	1860	3380
1021	1870	3398
1027	1880	3416
1032	1890	3434
1038	1900	3452
1043	1910	3470
1049	1920	3488
1054	1930	3506
1060	1940	3524
1066	1950	3542
1071	1960	3560
1077	1970	3578
1082	1980	3596
1088	1990	3614
1093	2000	3632
1099	2010	3650
1104	2020	3668
1110	2030	3686
1116	2040	3704
1121	2050	3722
1127	2060	3740
1132	2070	3758
1138	2080	3776
1143	2090	3794
1149	2100	3812
1154	2110	3830
1160	2120	3848
1166	2130	3866
1171	2140	3884
1177	2150	3902
1182	2160	3920
1188	2170	3938
1193	2180	3956
1199	2190	3974
1204	2200	3992
1210	2210	4010
1216	2220	4028
1221	2230	4046
1227	2240	4064
1232	2250	4082
1238	2260	4100

2270 to 3000

C	(value)	F
1243	2270	4118
1249	2280	4136
1254	2290	4154
1260	2300	4172
1266	2310	4190
1271	2320	4208
1277	2330	4226
1282	2340	4244
1288	2350	4262
1293	2360	4280
1299	2370	4298
1304	2380	4316
1310	2390	4334
1316	2400	4352
1321	2410	4370
1327	2420	4388
1332	2430	4406
1338	2440	4424
1343	2450	4442
1349	2460	4460
1354	2470	4478
1360	2480	4496
1366	2490	4514
1371	2500	4532
1377	2510	4550
1382	2520	4568
1388	2530	4586
1393	2540	4604
1399	2550	4622
1404	2560	4640
1410	2570	4658
1416	2580	4676
1421	2590	4694
1427	2600	4712
1432	2610	4730
1438	2620	4748
1443	2630	4766
1449	2640	4784
1454	2650	4802
1460	2660	4820
1466	2670	4838
1471	2680	4856
1477	2690	4874
1482	2700	4892
1488	2710	4910
1493	2720	4928
1499	2730	4946
1504	2740	4964
1510	2750	4982
1516	2760	5000
1521	2770	5018
1527	2780	5036
1532	2790	5054
1538	2800	5072
1543	2810	5090
1549	2820	5108
1554	2830	5126
1560	2840	5144
1566	2850	5162
1571	2860	5180
1577	2870	5198
1582	2880	5216
1588	2890	5234
1593	2900	5252
1599	2910	5270
1604	2920	5288
1610	2930	5306
1616	2940	5324
1621	2950	5342
1627	2960	5360
1632	2970	5378
1638	2980	5396
1643	2990	5414
1649	3000	5432

SPEED-O-GRAPH

REFRIGERANT PIPE SIZE SELECTOR

For rapid estimating of refrigerant line sizes for average air-conditioning applications based on 40°F suction and 105°F condensing temperatures at average pressure loss 2°F.

R-12

EQUIVALENT LENGTH OF PIPE IN FEET

TONS	20 FT.			30 FT.			40 FT.			50 FT.			60 FT.			70 FT.			80 FT.			90 FT.			100 FT.		
	LIQ.	SUCT.	DISCH.	LIQ.	SUCT.	DISCH.	LIQ.	SUCT.	DISCH.	LIQ.	SUCT.	DISCH.	LIQ.	SUCT.	DISCH.	LIQ.	SUCT.	DISCH.	LIQ.	SUCT.	DISCH.	LIQ.	SUCT.	DISCH.	LIQ.	SUCT.	DISCH.
2	1/2	3/4	5/8	1/2	7/8	5/8	1/2	7/8	3/4	1/2	7/8	3/4	1/2	1 1/8	3/4	1/2	1 1/8	3/4	1/2	1 1/8	3/4	1/2	1 1/8	3/4	1/2	1 1/8	3/4
3	1/2	7/8	3/4	1/2	1 1/8	3/4	1/2	1 1/8	3/4	1/2	1 1/8	3/4	5/8	1 1/8	3/4	5/8	1 1/8	7/8	5/8	1 1/8	7/8	5/8	1 1/8	7/8	5/8	1 1/8	7/8
5	1/2	1 1/8	7/8	5/8	1 1/8	7/8	5/8	1 1/8	7/8	5/8	1 3/8	1 1/8	5/8	1 3/8	1 1/8	3/4	1 3/8	1 1/8	5/8	1 3/8	1 1/8	3/4	1 3/8	1 1/8	3/4	1 3/8	1 1/8
8	5/8	1 3/8	1 1/8	5/8	1 3/8	1 1/8	3/4	1 3/8	1 1/8	3/4	1 5/8	1 1/8	3/4	1 5/8	1 1/8	3/4	1 5/8	1 3/8	7/8	1 5/8	1 3/8	7/8	1 5/8	1 3/8	7/8	2 1/8	1 3/8
10	5/8	1 3/8	1 1/8	3/4	1 5/8	1 1/8	3/4	1 5/8	1 1/8	3/4	1 5/8	1 3/8	7/8	1 5/8	1 3/8	7/8	1 5/8	1 3/8	7/8	2 1/8	1 3/8	7/8	2 1/8	1 3/8	7/8	2 1/8	1 3/8
15	3/4	1 5/8	1 3/8	7/8	1 5/8	1 3/8	7/8	2 1/8	1 3/8	7/8	2 1/8	1 3/8	1 1/8	2 1/8	1 5/8	1 1/8	2 1/8	1 5/8	1 1/8	2 1/8	1 5/8	1 1/8	2 1/8	1 5/8	1 1/8	2 1/8	1 5/8
20	7/8	1 5/8	1 3/8	7/8	2 1/8	1 3/8	1 1/8	2 1/8	1 5/8	1 1/8	2 1/8	1 5/8	1 1/8	2 1/8	1 5/8	1 1/8	2 1/8	1 5/8	1 1/8	2 5/8	2 1/8	1 1/8	2 5/8	2 1/8	1 1/8	2 5/8	2 1/8
25	7/8	2 1/8	1 5/8	1 1/8	2 1/8	1 5/8	1 1/8	2 1/8	1 5/8	1 1/8	2 5/8	1 5/8	1 1/8	2 5/8	2 1/8	1 1/8	2 5/8	2 1/8	1 1/8	2 5/8	2 1/8	1 3/8	2 5/8	2 1/8	1 3/8	2 5/8	2 1/8
30	1 1/8	2 1/8	1 5/8	1 1/8	2 1/8	1 5/8	1 1/8	2 5/8	2 1/8	1 1/8	2 5/8	2 1/8	1 1/8	2 5/8	2 1/8	1 3/8	2 5/8	2 1/8	1 3/8	2 5/8	2 1/8	1 3/8	2 5/8	2 1/8	1 3/8	2 5/8	2 1/8
40	1 1/8	2 1/8	1 5/8	1 1/8	2 5/8	2 1/8	1 3/8	2 5/8	2 1/8	1 3/8	2 5/8	2 1/8	1 3/8	2 5/8	2 1/8	1 3/8	3 1/8	2 1/8	1 3/8	3 1/8	2 1/8	1 3/8	3 1/8	2 5/8	1 5/8	3 1/8	2 5/8
50	1 1/8	2 5/8	2 1/8	1 3/8	2 5/8	2 1/8	1 3/8	2 5/8	2 1/8	1 3/8	3 1/8	2 1/8	1 3/8	3 1/8	2 5/8	1 5/8	3 1/8	2 5/8	1 5/8	3 1/8	2 5/8	1 5/8	3 5/8	2 5/8	1 5/8	3 5/8	2 5/8
60	1 3/8	2 5/8	2 1/8	1 3/8	3 1/8	2 1/8	1 3/8	3 1/8	2 5/8	1 5/8	3 1/8	2 5/8	1 5/8	3 1/8	2 5/8	1 5/8	3 5/8	2 5/8	1 5/8	3 5/8	2 5/8	1 5/8	3 5/8	2 5/8	1 5/8	3 5/8	2 5/8
70	1 3/8	2 5/8	2 1/8	1 3/8	3 1/8	2 5/8	1 5/8	3 1/8	2 5/8	1 5/8	3 5/8	2 5/8	1 5/8	3 5/8	2 5/8	1 5/8	3 5/8	2 5/8	2 1/8	3 5/8	2 5/8	2 1/8	3 5/8	3 1/8	2 1/8	4 1/8	3 1/8
80	1 3/8	3 1/8	2 1/8	1 5/8	3 1/8	2 5/8	1 5/8	3 5/8	2 5/8	1 5/8	3 5/8	2 5/8	1 5/8	3 5/8	2 5/8	2 1/8	3 5/8	3 1/8	2 1/8	4 1/8	3 1/8	2 1/8	4 1/8	3 1/8	2 1/8	4 1/8	3 1/8
90	1 3/8	3 1/8	2 5/8	1 5/8	3 1/8	2 5/8	1 5/8	3 5/8	2 5/8	1 5/8	3 5/8	2 5/8	2 1/8	3 5/8	3 1/8	2 1/8	4 1/8	3 1/8	2 1/8	4 1/8	3 1/8	2 1/8	4 1/8	3 1/8	2 1/8	4 1/8	3 1/8
100	1 5/8	3 1/8	2 5/8	1 5/8	3 5/8	2 5/8	1 5/8	3 5/8	2 5/8	2 1/8	3 5/8	3 1/8	2 1/8	4 1/8	3 1/8	2 1/8	4 1/8	3 1/8	2 1/8	4 1/8	3 1/8	2 1/8	4 1/8	3 1/8	2 1/8	5 1/8	3 1/8

CONDENSATE LINE SIZING, CONDENSER TO RECEIVER

CONDENSATE LINE SIZE	1/2	5/8	3/4	7/8	1 1/8	1 3/8	1 5/8	2 1/8	2 5/8	3 1/8	3 5/8	4 1/8
REFRIGERATION MAXIMUM TONS	1.5	3.0	5.0	7.5	14.0	23.0	35.0	70.0	110.0	170.0	245.0	330.0
MINIMUM HEIGHT	10''	10''	10''	10''	14''	14''	14''	16''	16''	16''	16''	16''

R-22

EQUIVALENT LENGTH OF PIPE IN FEET

TONS	20 FT.			30 FT.			40 FT.			50 FT.			60 FT.			70 FT.			80 FT.			90 FT.			100 FT.		
	LIQ.	SUCT.	DISCH.	LIQ.	SUCT.	DISCH.	LIQ.	SUCT.	DISCH.	LIQ.	SUCT.	DISCH.	LIQ.	SUCT.	DISCH.	LIQ.	SUCT.	DISCH.	LIQ.	SUCT.	DISCH.	LIQ.	SUCT.	DISCH.	LIQ.	SUCT.	DISCH.
2	3/8	5/8	1/2	3/8	5/8	1/2	3/8	3/4	1/2	3/8	3/4	5/8	3/8	3/4	5/8	1/2	3/4	5/8	1/2	3/4	5/8	1/2	3/4	5/8	1/2	7/8	5/8
3	3/8	3/4	5/8	3/8	3/4	5/8	1/2	3/4	5/8	1/2	7/8	5/8	1/2	7/8	5/8	1/2	7/8	3/4	1/2	7/8	3/4	1/2	7/8	3/4	1/2	7/8	3/4
5	1/2	7/8	5/8	1/2	7/8	5/8	1/2	1 1/8	3/4	1/2	1 1/8	3/4	5/8	1 1/8	3/4	5/8	1 1/8	7/8	5/8	1 1/8	7/8	5/8	1 1/8	7/8	5/8	1 1/8	7/8
8	1/2	1 1/8	3/4	5/8	1 1/8	3/4	5/8	1 1/8	7/8	5/8	1 1/8	7/8	5/8	1 1/8	1 1/8	5/8	1 3/8	1 1/8	3/4	1 3/8	1 1/8	3/4	1 3/8	1 1/8	3/4	1 3/8	1 1/8
10	5/8	1 1/8	7/8	5/8	1 1/8	3/4	5/8	1 1/8	7/8	5/8	1 3/8	1 1/8	3/4	1 3/8	1 1/8	3/4	1 3/8	1 1/8	3/4	1 3/8	1 1/8	3/4	1 3/8	1 1/8	3/4	1 3/8	1 1/8
15	5/8	1 3/8	1 1/8	3/4	1 3/8	1 1/8	3/4	1 3/8	1 1/8	3/4	1 3/8	1 1/8	3/4	1 5/8	1 1/8	7/8	1 5/8	1 1/8	7/8	1 5/8	1 3/8	7/8	1 5/8	1 3/8	7/8	1 5/8	1 3/8
20	3/4	1 3/8	1 1/8	3/4	1 3/8	1 1/8	7/8	1 5/8	1 1/8	7/8	1 5/8	1 3/8	7/8	1 5/8	1 3/8	7/8	1 5/8	1 3/8	7/8	2 1/8	1 3/8	1 1/8	2 1/8	1 3/8	1 1/8	2 1/8	1 3/8
25	3/4	1 5/8	1 1/8	7/8	1 5/8	1 1/8	7/8	1 5/8	1 3/8	7/8	2 1/8	1 3/8	1 1/8	2 1/8	1 3/8	1 1/8	2 1/8	1 3/8	1 1/8	2 1/8	1 3/8	1 1/8	2 1/8	1 5/8	1 1/8	2 1/8	1 5/8
30	7/8	1 5/8	1 3/8	7/8	1 5/8	1 3/8	7/8	2 1/8	1 3/8	1 1/8	2 1/8	1 3/8	1 1/8	2 1/8	1 5/8	1 1/8	2 1/8	1 5/8	1 1/8	2 1/8	1 5/8	1 1/8	2 1/8	1 5/8	1 1/8	2 1/8	1 5/8
40	7/8	2 1/8	1 3/8	1 1/8	2 1/8	1 3/8	1 1/8	2 1/8	1 5/8	1 1/8	2 1/8	1 5/8	1 1/8	2 1/8	1 5/8	1 1/8	2 1/8	1 5/8	1 1/8	2 5/8	1 5/8	1 3/8	2 5/8	2 1/8	1 3/8	2 5/8	2 1/8
50	1 1/8	2 1/8	1 3/8	1 1/8	2 1/8	1 5/8	1 1/8	2 1/8	1 5/8	1 1/8	2 5/8	1 5/8	1 3/8	2 5/8	2 1/8	1 3/8	2 5/8	2 1/8	1 3/8	2 5/8	2 1/8	1 3/8	2 5/8	2 1/8	1 3/8	2 5/8	2 1/8
60	1 1/8	2 1/8	1 5/8	1 1/8	2 1/8	1 5/8	1 1/8	2 5/8	2 1/8	1 3/8	2 5/8	2 1/8	1 3/8	2 5/8	2 1/8	1 3/8	2 5/8	2 1/8	1 3/8	2 5/8	2 1/8	1 3/8	2 5/8	2 1/8	1 3/8	3 1/8	2 1/8
70	1 1/8	2 1/8	1 5/8	1 1/8	2 5/8	2 1/8	1 3/8	2 5/8	2 1/8	1 3/8	2 5/8	2 1/8	1 3/8	2 5/8	2 1/8	1 3/8	2 5/8	2 1/8	1 3/8	3 1/8	2 1/8	1 5/8	3 1/8	2 1/8	1 5/8	3 1/8	2 1/8
80	1 1/8	2 5/8	1 5/8	1 3/8	2 5/8	2 1/8	1 3/8	2 5/8	2 1/8	1 3/8	2 5/8	2 1/8	1 3/8	3 1/8	2 1/8	1 5/8	3 1/8	2 1/8	1 5/8	3 1/8	2 1/8	1 5/8	3 1/8	2 5/8	1 5/8	3 1/8	2 5/8
90	1 1/8	2 5/8	2 1/8	1 3/8	2 5/8	2 1/8	1 3/8	2 5/8	2 1/8	1 3/8	3 1/8	2 1/8	1 5/8	3 1/8	2 1/8	1 5/8	3 1/8	2 5/8	1 5/8	3 1/8	2 5/8	1 5/8	3 1/8	2 5/8	1 5/8	3 1/8	2 5/8
100	1 3/8	2 5/8	2 1/8	1 3/8	2 5/8	2 1/8	1 3/8	3 1/8	2 1/8	1 5/8	3 1/8	2 1/8	1 5/8	3 1/8	2 5/8	1 5/8	3 1/8	2 5/8	1 5/8	3 1/8	2 5/8	1 5/8	3 1/8	2 5/8	1 5/8	3 5/8	2 5/8

CONDENSATE LINE SIZING, CONDENSER TO RECEIVER

CONDENSATE LINE SIZE	1/2	5/8	3/4	7/8	1 1/8	1 3/8	1 5/8	2 1/8	2 5/8	3 1/8	3 5/8	4 1/8
REFRIGERATION MAXIMUM TONS	1.8	3.7	6.0	9.0	17.0	28.0	42.0	84.0	131.0	205.0	295.0	395.0
MINIMUM HEIGHT	10''	10''	10''	10''	14''	14''	14''	16''	16''	16''	16''	16''

FIGURE A1

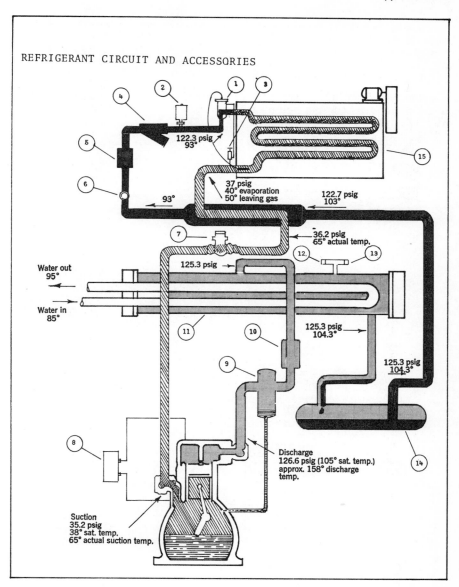

FIGURE A2

REFRIGERANT CIRCUIT AND ACCESSORIES

Most examinations will have one or more questions related to the *refrigeration circuit* and the location of accessories in the circuit. The candidate should have a good graphic concept of such a circuit in his minds eye and be thoroughly familiar with the change of state of the refrigerant within the circuit.

Figure 9.10 diagrams a *refrigeration circuit* with the standard accessories and shows the change of state between liquid and gas for a system using R-12.

▨▨▨	Low pressure gas
▨▨▨	Low pressure liquid
▨▨▨	High pressure gas
▨▨▨	High pressure liquid
▨▨▨	Oil
①	Expansion valve
②	Solenoid valve
③	External equalizer
④	Strainer
⑤	Dehydrator
⑥	Sight glass
⑦	Evaporator pressure regulator
⑧	Hi-lo cut-out
⑨	Oil separator
⑩	Muffler
⑪	Condenser
⑫	Vent valve
⑬	Relief valve
⑭	Liquid receiver
⑮	Evaporator

INDEX

NOTES

NOTES